PUZZLES IN
MATH AND
LOGIC

PUZZLES IN MATH AND LOGIC

100 NEW RECREATIONS

BY

AARON J.
FRIEDLAND

DOVER PUBLICATIONS, INC.

NEW YORK

Published in Canada by General Publishing Company, Ltd., 30 Lesmill Road, Don Mills, Toronto,
Ontario.
Published in the United Kingdom by Constable
and Company, Ltd., 10 Orange Street, London WC 2.

Puzzles in Math and Logic is a new work, first
published by Dover Publications, Inc., in 1970.

International Standard Book Number: 0–486–22256–X
Library of Congress Catalog Card Number: 72–130639

Manufactured in the United States of America
Dover Publications, Inc.
180 Varick Street
New York, N.Y. 10014

Preface

As you thumb through this volume, you may smile with pleasure as you come across your favorite puzzle. However, I sincerely hope that you will be deprived of this experience, as I have attempted to make this a collection of original puzzles.

There are various ways of classifying mathematical puzzles. They may be grouped according to the type of mathematics needed to solve them. Thus, this collection may be divided into problems in logic such as Nos. 38 and 94, problems involving probability such as Nos. 10 and 79, problems dealing with the properties of numbers such as Nos. 4 and 53, geometric dissections such as Nos. 5 and 24, counter arrangements such as Nos. 67 and 81, and so on over a broad range of categories in formal and recreational mathematics. The puzzles range in difficulty from the fairly involved logical structure of No. 94 or the combinational complexities of No. 39, down to the fairly simple reasoning of No. 91 and straightforward calculations of No. 9, with a few "trick" questions thrown in for good measure. In general they are reasonably difficult and intended to appeal to the serious puzzle solver. Although not all of the puzzles may be to everyone's taste, there are enough different kinds of puzzles on different levels to provide something for everyone.

I prefer to classify mathematical puzzles in a manner which does not depend upon the branch of mathematics involved. The elements (not mutually exclusive) which I seek in a puzzle are "challenge" and "surprise." No. 39 is a good example of a challenge puzzle. There is no formal procedure for arriving at the solution, but by a combination of insight and perseverance one can solve the puzzle and feel a sense of accomplishment.

The element of surprise may occur in the method of solution of a puzzle. Here the enjoyment lies in discovering an unexpectedly simple method of solving an apparently difficult problem. I do not think I am revealing too much by stating that No. 2 is an example of such a puzzle with a surprise approach.

Perhaps of greatest appeal is the puzzle with a surprise result. In its ideal form, the solution is paradoxical or violates common sense. Sometimes it is apparent from the statement of the puzzle that it belongs in this category. In No. 1, for example, the apparent solution is much too obvious to be correct. However, any other solution appears to violate the rules of the game.

In the pages ahead, I hope you will find a number of challenges and a number of surprises.

Contents

PUZZLES IN
MATH AND
LOGIC

1

POKER HANDS

Which of the following poker hands is the best? Which is the worst? Which hands are of equal strength? The game is being played with an ordinary 52-card pack. There are no wild cards. (AS means ace of spades, etc.)

(a) AS AH AD KS KH
(b) AS AH AD QS QC
(c) AS AH AD QS QH
(d) AS AH AD 6S 6C
(e) AS AH AD 3S 3C

2

100 DIGITS

100 digits are chosen at random. What is the probability that there will be at least one pair of numbers, a and b, such that each digit is used exactly once in forming the two numbers, and such that $a^2 = b$? Of course, no number may begin with the digit 0.

3

THE CHESS MATCH

Abner Appleby is scheduled to play a two game chess match with Barney Blitz. If they are tied after two games, there will be a play-off, and the first player to win a game thereafter wins the match. The situation does not look too promising for Abner, since Barney is the stronger player. Abner can play a daring game, which he has a 45% probability of winning and a 55% probability of losing; or he can play a conservative game, which he has a 90% probability of drawing and a 10% probability of losing. Either way he would wind up ten games behind out of every hundred played. What is Abner's best strategy, and what is the probability of his winning the match?

4

THE PROFESSOR'S TELEPHONE NUMBER

"How is it, Professor Flugel," asked the puzzled student, "that one so notoriously absentminded as yourself manages to remember his telephone number?"

"Quite simple, young man," replied the professor. "I simply keep in mind that my telephone number is the only seven-digit number which is converted into a factor of itself when the order of its digits is reversed."

What is the professor's telephone number? Are there any other numbers which have this characteristic, excluding, of course, the trivial cases of numbers which read the same forward and backward?

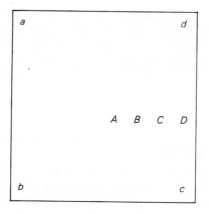

5

HOUSES AND WELLS

Old Zeke left his property to his four sons with instructions that Abner was to get house *A* and well *a*, Barney was to get house *B* and well *b*, Cyrus was to get house *C* and well *c*, and Dagwood was to get house *D* and well *d*. Furthermore, each of the four pieces of land containing a house and a well was to be of the same size and shape. How was the land divided?

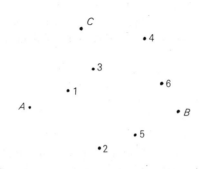

6

FOUR POINTS

Billy Jones found that with the three points *A*, *B*, and *C* arranged as in the above sketch, he could place a fourth point where it would be closest to any one of the three, next closest to either of the remaining two, and furthest from the remaining one. Points 1 through 6 illustrate each possible combination. He ran into some difficulty when he tried to arrange four points in a similar manner, so that a fifth point could be placed where it would be closest to any one of the four, next closest to any of the remaining three, etc. However, he finally succeeded. How did he arrange the four points?

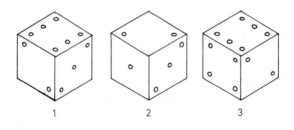

7

THE CASE OF THE CURIOUS CUBE

Here are three views of the same cube. How many spots are there on the bottom face in view 1 (the face opposite the "six")?

8

HEXAGONS

When Henry cut two similar irregular hexagons out of a sheet of paper, he was surprised to find that he could not completely cover the smaller one with the larger. What were their shapes?

9

AN UNUSUAL NUMBER

Wilfred Stump heard that there is only one number between 2 and 200,000,000,000,000 which is a perfect square, cube, and fifth power, and has decided to look for it. So far he has checked out every number up to about 100,000 and is beginning to get somewhat discouraged. Perhaps the reader can help him out.

10

TOSSING COINS

(a) Abner, Boris, and Charlie toss a coin 15, 16, and 17 times, respectively. Which one is least likely to have tossed more heads than tails?

(b) The same, except that the coins are tossed 15, 17, and 20 times, respectively.

(c) The same, except that the coins are tossed 18, 19, and 20 times, respectively.

11

THE CHESS GAME

"Charlie Brown is a sucker for punishment," said Bert to Marge. "I played him a game of chess this morning and had him hopelessly beaten, but he refused to resign and insisted that I mate him. Look at this photograph I took of the chessboard, and you can see how ridiculous it was for him to continue. I had white, of course."

"It was my move," Bert continued, "so naturally I mated him on that move."

How did white mate in one move?

12

FATHER AND SON

If Hubert's age is multiplied by the age of his father, the product is a permutation of the digits in their individual ages. How old are they?

13

THE GAMBLER

"See that gray-haired chap betting red or black at the roulette table?" said George. "Loves to gamble, but he ran afoul of a witch doctor who put a curse on him. Now he loses consistently about 80% of the time. Drops about $50,000 a year."

"He must be pretty wealthy to be able to afford it," said Lennie.

"Not really. His wife supports him. That attractive woman at his right."

"I'm surprised she doesn't get fed up and leave him."

"On the contrary, she's quite happy about it."

Can you hazard a guess as to why the woman should be so pleased about her husband's rotten luck?

14

JOHN'S COUSINS

"This might interest you, professor," said John. "My age and the ages of each of my three distant cousins are all prime numbers, and the sum of our ages is 50."

"In that case," said the professor, who knew John's age, "I can tell you the ages of your three cousins."

You do not share the professor's advantage of knowing John's age to start with, but nevertheless, can you tell the ages of his cousins? (Note that the number 1 is not considered to be a prime).

MULTIPLICATIONS

In Problem 15 through 23, each of the ten digits in the multiplicand, multiplier, and product is different. For all but the last problem, one of the digits is supplied, and it is up to the reader to fill in the remaining ones.

15
$$
\begin{array}{r}
\text{xxx} \\
\underline{\text{xx}} \\
\text{xxxx1}
\end{array}
$$

16
$$
\begin{array}{r}
\text{xx2} \\
\underline{\text{xx}} \\
\text{xxxxx}
\end{array}
$$

17
$$
\begin{array}{r}
\text{xxx} \\
\underline{\text{xx}} \\
\text{3xxxx}
\end{array}
$$

18
$$
\begin{array}{r}
\text{xxx} \\
\underline{\text{x4}} \\
\text{xxxxx}
\end{array}
$$

19
$$\begin{array}{r} 5\text{xx} \\ \text{xx} \\ \hline \text{xxxxx} \end{array}$$

20
$$\begin{array}{r} \text{xx6} \\ \text{xx} \\ \hline \text{xxxxx} \end{array}$$

21
$$\begin{array}{r} \text{xxx} \\ 7\text{x} \\ \hline \text{xxxxx} \end{array}$$

22
$$\begin{array}{r} \text{xxx} \\ \text{xx} \\ \hline \text{xxx8x} \end{array}$$

23

Find another solution which is not based upon any of the given information in Problems 15 through 22. That is, a solution in which the last digit in the bottom row is not 1 (because such is given in Problem 15), the last digit in the top row is not 2 (because such is given in Problem 16), etc.

24

DISSECTING THE CHESSBOARD

A chessboard may easily be divided into four pieces of the same size and shape, such that each piece contains the same number of black and white squares. However, the reader is asked to perform the dissection in such a way that each piece contains more than twice as many squares of one color than of the other. The cuts are to be made in the usual manner along the edges of the squares and not cutting through any square.

25

FOUL PLAY

The Detroit Tigers and the Boston Red Sox in the Eastern Division of the American League each have 100 wins, 61 losses, and one game left to play. Yet Boston can clinch the division championship during the regular season, while Detroit at best can wind up in a tie for first place. How do you account for this obviously unfair situation?

26

WEATHER REPORT

"Can you tell me what the temperature has been at noon for the past five days?" John asked the weatherman.

"I don't exactly recall," replied the weatherman, "but I do remember that the temperature was different each day, and that the product of the temperatures is 12."

Assuming that the temperatures are expressed to the nearest degree, what were the five temperatures?

27

WHAT PRICE LUNCH?

"Shall we flip for the lunch checks?" asked Professor Wormhole on the way to the university cafeteria.

"I'll tell you what," said Professor Flugel. "If your check comes to exactly $1.15, I'll pay it. If not, you pick up the checks."

"Agreed," said Professor Wormhole.

However, although food items at all prices were available, and although Professor Wormhole tried his best to be billed for $1.15, nevertheless he was stuck with the checks. Can you suggest a reason?

The situation is one which is frequently encountered, although the amount of money involved may vary.

28

DIGITS EQUAL POWER

What is the largest n-digit number which is also an exact nth power?

29

ADDITIONS

The letters a through i stand for the digits 1 through 9 in the following two additions. Determine which letter stands for which digit. There are two solutions.

$$\begin{array}{r} abc \\ \underline{def} \\ ghi \end{array} \qquad \begin{array}{r} adg \\ \underline{beh} \\ cfi \end{array}$$

30

ADDITIONAL ADDITIONS

The letters a through i stand for the digits 1 through 9 in the following two additions. Determine which letter stands for which digit.

$$\begin{array}{r} abc \\ \underline{def} \\ ghi \end{array} \qquad \begin{array}{r} gda \\ \underline{heb} \\ ifc \end{array}$$

31

SQUARE PATTERNS

There are 16 positions that a 2 × 2 square can occupy within a 5 × 5 array of squares. There are also 16 possible patterns of black and white squares in a 2 × 2 array. This led Barney Boob to wonder if it were possible to color the squares in a 5 × 5 array black or white, so that each possible 2 × 2 pattern would be represented. Sammy Sharp has bet him he cannot do it. While

Barney is working away, Sammy is chuckling to himself because he realizes that the 16 2 × 2 arrays must have the same total number of black and white squares, while the 5 × 5 array must have a different number of each color. Should Barney give up, in the face of this evidence that his quest is impossible?

32

A BAG OF MARBLES

The Miracle Marble Manufacturing Company manufactures orange marbles and purple marbles. A bag of their marbles may contain any combination of orange and purple marbles (including all orange or all purple) and all combinations are equally probable. Henry bought a bag of their marbles and pulled one out at random. It was purple. What is the probability that if he pulled out a second marble at random it would also be purple?

33

ELLIPSE

In Ho Ming's garden, there is an elliptical pond whose area is exactly fifty square feet. Along the border of the pond is a flower bed two feet wide, and bordering the flower bed is a path also two feet wide. The entire region may be assumed to lie in a plane.

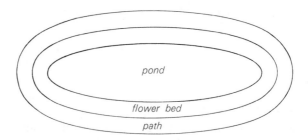

When Ho Ming's guest, Lo Fat, remarked on the aesthetically pleasing curves formed by the three boundary lines, Ho Ming described the arrangement to him and then added, "Perhaps you can tell me what is the area enclosed by the largest ellipse?"

Lo Fat, well aware of his host's reputation for trickery, pondered for a few minutes and then gave the correct answer. What was it?

1   2   3

34

ANOTHER CURIOUS CUBE

Here are three views of the same cube. Each letter stands for a different color. What color is the bottom face in Figure 1 (the face opposite the one colored A)?

35

THE RACE FOR THE DIVISION CHAMPIONSHIP

Boston and Detroit are the only remaining contenders for the championship. Each has won 2/3 of its games. However, Boston has completed its schedule, while Detroit has three games left. If Detroit were to win two and lose one, it would finish in a tie with Boston, and have a 50% probability of winning the play-off and the championship. However, the team it is playing is slightly better than average, and instead of a 2/3 ($66\frac{2}{3}$%) probability of winning, Detroit has only a 66% probability of winning each game. As a result, the probability of Detroit winning the championship is about

(a) 45%   (b) 47%   (c) 49%   (d) 51%

36

A GROUP OF NUMBERS

Find the largest group of different positive integers less than 100 such that no combination of them added together totals 100.

37

TWO ADDITIONS

The following two addition examples are correct as written. They are also correct if a different digit is substituted for each letter in the examples. Find the proper substitutions.

$$\begin{array}{r} \text{ONE} \\ + \text{ONE} \\ \hline \text{TWO} \end{array} \qquad\qquad \begin{array}{r} \text{ONE} \\ + \text{FOUR} \\ \hline \text{FIVE} \end{array}$$

38

REGARDING TRUTH AND LIES

During his sabbatical, Professor Flugel visited that favorite puzzleland country in which there are only two types of inhabitants, those who always speak the truth and those who always lie. They also always answer "Yes" or "No" to any question for which such an answer is meaningful. During his visit, the professor amused himself by asking a question which none of the inhabitants could answer, although the question could be answered "Yes" or "No" and did not require any factual knowledge of which the inhabitants were unaware. What was the question?

39

FIVE WEIGHTS

A chemist has a set of five weights. He knows that it includes one 1-gram weight, and also one each 2-, 3-, 4-, and 5-gram weights, but because they are unmarked he has no way of telling them apart except by placing them on a balance. He may place any combination of weights on each of the two pans and determine if one side is heavier than the other or if they balance. Show how in five weighings he can identify each of the weights.

40

FEWEST WEIGHINGS

Prove that the five weights in Problem 39 cannot be identified in fewer than five weighings.

MORE ADDITIONS

Problems 41 through 51 are addition problems in which each of the ten digits is different. In each column of digits to be added, the digits should be arranged in increasing order from the top to the bottom of the column.

41
```
 3xx
 xxx
────
xxxx
```

42
```
 x3x
 xxx
────
xxxx
```

43
```
 xxx
 xxx
────
x3xx
```

44
```
 xxx
 xxx
────
xx3x
```

45
```
 6xx
 xxx
────
xxxx
```

46
```
 x6x
 xxx
────
xxxx
```

47

```
    xxx
    xxx
   ------
   x6xx
```

48

```
    xxx
    xxx
   ------
   xx6x
```

49

```
      4
     xx
    xxx
   ------
   xxxx
```

50

```
      x
     4x
    xxx
   ------
   xxxx
```

51

```
      x
     x4
    xxx
   ------
   xxxx
```

52

ANOTHER CHESS GAME

"I saw you playing chess with Charlie Brown again," said Marge to Bert. "I suppose he played his usual miserable game."

"That's right," said Bert. "This was our position toward the end of the game. It was my move."

Marge studied the board for several minutes. Then she commented, "I suppose you mated him on that move."

"Naturally," replied Bert.

What was the move?

BLACK: Charlie Brown

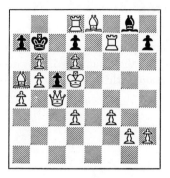

WHITE: Bert

53

THE PERFECT SQUARE

Little Suzy arranged her numbered blocks to form five numbers, one of which was a perfect square. She then went to fetch her father, to see how long it would take him to figure out which number was the square. When they returned, however, they discovered that Suzy's younger brother, Winston, had rearranged the blocks in the first number and in the second number, and had walked off with all but the last two blocks in the third number, all but the last block in the fourth number, and all of the blocks in the fifth number. This is what the remaining blocks looked like:

(1)	3	5	5	8	7	3	2
(2)	3	0	8	7	8	2	1
(3)						7	5
(4)							8
(5)							

Suzy was crest-fallen, but after looking at the blocks for several moments her father said, "Never mind, dear, I can tell you which number was the square." How did he know?

54

OBJECT IN MOTION

An object is in motion through a stationary medium. If the surrounding medium were absent, the object would continue moving indefinitely at the same speed. However, because of the energy lost as friction with the medium, the object speeds up. Explain.

55

ANGLES

"Professor Flugel, I've just constructed these two angles," said Archie. "I can prove that angle B is exactly $\frac{1}{3}$ of angle A. Yet I performed the construction in a finite number of steps using only an unmarked straight-edge and a compass. Furthermore, angle A does not have some special value, such as 45°, which would make the construction possible."

"I believe you are mistaken," said Professor Flugel, who was well aware that it is impossible to trisect an angle under the conditions stated. However, when Archie demonstrated his method, Professor Flugel was chagrined to find that Archie had accomplished exactly what he had claimed.

How was the construction performed?

56

THE BETTER HAND

Which of the following two poker hands is stronger? The game is being played with an ordinary 52-card pack and there are no wild cards.

(a) AS AH AD KS KH
(b) AS AH KS KD KC

57

PYTHAGOREAN PARALLELEPIPED

You are familiar with Pythagorean triangles—integral solutions of the equation $x^2 + y^2 = z^2$—such as 3, 4, 5; 5, 12, 13; etc. The problem now is to find a three-dimensional example—a rectangular parallelepiped whose edges and diagonal can be expressed as integers—i.e., an integral solution of $w^2 + x^2 + y^2 = z^2$. The problem should be solved mentally.

58

THE TWO-TONE CUBE

"I have here a mysterious cube," said Sammy Sharp. "Notice that each face is half black and half white, and all are identical. Now if you will take the cube and hold it so that I can only see one face, I will tell you which half is black and which is white on each of the other five faces."

"Ha," said Barney Boob. "I can color a cube so that I can do the same trick." And he proceeded to do so. But oddly enough, his cube was different from Sammy's. In fact, they were not even mirror images of each other. What did the two cubes look like?

59

THE BASEBALL SEASON

The baseball season is approaching its close with New York, Detroit, and Washington in first, second, and third place, respectively. New York is very interested in the outcome of the game being played between Detroit and Washington, because if Detroit wins, New York has clinched the championship; while if Detroit loses, New York can still lose the championship. Why is this?

60

ROULETTE

Lenny came bursting into the room. "I've done it, George! I've developed the perfect system for winning at roulette."

George was skeptical. "I've gone broke on your systems before."

"No, really," said Lenny. "Look, I'll demonstrate. We'll take eighteen red and eighteen black cards out of this pack, to represent the roulette numbers. We'll only play the even money bets, red or black, so the values don't matter. Let's also put in two jokers. That's the house percentage, the zero and double-zero which also lose. Now, the system is to assume that the house has a certain amount of money to start with, say $100, and to always bet 20% of the house's money on red. That means you bet $20 to start with. Then if you win on the first play, you bet $16 the second time; while if you lose on the first play, you bet $24 the second time, and so on. On the average, reds, blacks, and zeros will come up as often in roulette as they do in this pack. Now, just shuffle the cards, run through the pack, and keep track of your bets on this sheet of paper."

Ten minutes later, George played the last card, and completed his calculation. "Well, I came out $30.94 ahead that time, but maybe it was just a lucky run of the cards."

"Shuffle the pack, and try it again," said Lenny.

Another ten minutes went by. George finished, and said with some surprise, "I came out $30.94 ahead this time too! Does it always work out this way?"

"Try running through the pack in the same order again," said Lenny, "only this time bet on black instead of red."

Ten minutes later, George was flabbergasted to get the same result. "I have to admit that you've really got something this time, Lenny. Come on, let's get down to Las Vegas and make a fortune."

The problem is, why is the author of this book so generously presenting this amazing system to the reader, instead of going to Las Vegas and making his own fortune?

61

THE NUMBER RING

Childless millionaire Linus Loot decided to leave his fortune to whichever one of his four nephews who could solve the following puzzle.

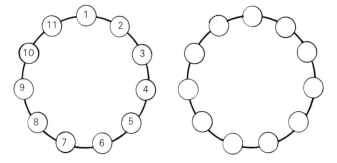

"I have numbered the circles in this ring from 1 to 11. Number the circles in the second ring from 1 to 11 in such an order that if the second ring is placed over the first in any position, right side up or inverted, at least one number will be at the same location in both rings."

Unfortunately, when they met again the following day, each nephew had come up with a different solution. Unable to decide which nephew to make his heir, Linus Loot took the sensible course and outlived them all.

What were the four solutions?

62

THE POKER GAME

Professor Flugel was kibitzing at a poker game. Walking around the table after the hands had been dealt and the betting was about to begin, he noted that the following hands were held:

John:	AS	KS	QS	JS	10S
Jim:	9S	9H	9D	9C	2S
Joe:	8S	8H	4S	4C	3D
Jason:	7S	7D	KH	QH	4D
Jerry:	7H	7C	KD	QC	4H

The players were all fairly good, and the usual rules were in effect. At this time, Max walked over to the professor and whispered, "Who has the worst hand?" What did the professor whisper back?

MORE MULTIPLICATIONS

In the following two multiplication problems, each of the ten digits is different.

63 $1 \times ab \times cde = fghi$

64 $2 \times ab \times cde = fghi$

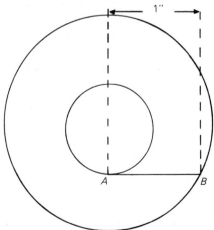

65

AREA OF ANNULUS

Line AB, which is one inch long, is tangent to the inner of two concentric circles at A and intersects the outer circle at B. What is the area of the annular region between the circles?

66

MORE TRUTH AND LIES

In a certain country, there are three types of inhabitants: type T who always tell the truth; type F who always lie; and type A who alternately lie and tell the truth. They will answer any question with "Yes" or "No" if one of these answers is possible. If the question cannot be answered, they will keep silent.

By asking an inhabitant one question, determine which type he is.

67

ROWS OF BUSHES

To celebrate her sixteenth birthday, Rose Busch decided to plant sixteen rose bushes in her garden. At first she considered planting them in a four-by-four square array, which would result in ten rows of four (four rows, four columns, and two diagonals). However, she finally decided that it would be more pleasing to plant the bushes in fifteen rows of four. How did she arrange them?

68

THE TENTH POWER

Wilfred Stump suspects that there is no ten–digit number in which all the digits are different, which is also a perfect tenth power. He has programmed a computer to check out every ten–digit number in order to verify this. What would be his best way of simplifying the amount of calculation required?

69

CARD TRICK

"I will now demonstrate a remarkable card trick," said Barney to his five friends. "Here is a pack of consecutively numbered cards arranged in order. I would like each of you in turn to give it a perfect riffle shuffle." (A perfect riffle shuffle consists of cutting the pack into halves, and combining them so that the cards in the halves alternate. The top card in what was formerly the lower half becomes the top card in the combined pack.)

After each of his five friends completed a perfect riffle shuffle, Barney turned up the cards one by one, and his friends were amazed to find that the pack was in the original order.

Barney then added two cards to the pack, repeated the trick, and again his friends were surprised at the result.

(a) How many cards were in the original pack?

(b) What were the results when two cards were added to the pack and the trick repeated?

STILL MORE MULTIPLICATIONS

In Problems 70 through 76 each of the digits from 1 through 9 appears exactly once in the multiplicand, multiplier, or product. In each case, one of the digits is given, and the problem is to fill in the remaining ones.

70
$$
\begin{array}{r}
2\text{xx} \\
\text{xx} \\
\hline
\text{xxxx}
\end{array}
$$

71
$$
\begin{array}{r}
\text{xx}3 \\
\text{xx} \\
\hline
\text{xxxx}
\end{array}
$$

72
$$
\begin{array}{r}
\text{x}3\text{x} \\
\text{xx} \\
\hline
\text{xxxx}
\end{array}
$$

73
$$\begin{array}{r} \text{xxx} \\ \underline{\text{xx}} \\ \text{4xxx} \end{array}$$

74
$$\begin{array}{r} \text{xxx} \\ \underline{\text{xx}} \\ \text{xx5x} \end{array}$$

75
$$\begin{array}{r} \text{xxx} \\ \underline{\text{xx}} \\ \text{x6xx} \end{array}$$

76
$$\begin{array}{r} \text{xxx} \\ \underline{\text{x7}} \\ \text{xxxx} \end{array}$$

77

THE DEPARTMENT STORE

Professor Flugel and his wife, Henrietta, were strolling through Bingle's Department Store. Henrietta stopped for a few minutes to examine the ladies' hat display, and she then discovered that she had become separated from her husband. She was about to look for him, when she recalled a lecture of his in which he had demonstrated that if two people are trying to find each other, it is more efficient if one stands still than if both search. This suited her and she decided to remain among the hats until her husband should find her.

However, in spite of the fact that her recollection regarding efficiency of searches was correct, it turned out that her strategy was unwise. Can you suggest why?

78

HALF A CHESSBOARD

Susan cut a chessboard in half, and wrote a digit in each square. The top eight-digit number was factorable by 3, the second by 5, the third by 7, and the bottom number by 11. She then left the room for several minutes. When she returned, she found that her younger brother Winston had cut out each of the squares. He had kept all eight digits of each number together, but the digits were no longer in their proper order, nor were the numbers themselves in the original order. The squares looked like this:

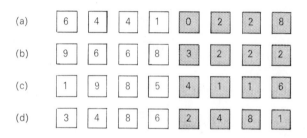

(a) 6 4 4 1 0 2 2 8

(b) 9 6 6 8 3 2 2 2

(c) 1 9 8 5 4 1 1 6

(d) 3 4 8 6 2 4 8 1

In what order were the rows originally arranged, and what was the number in the original third row (the number factorable by 7)?

79

THE CHESS TOURNAMENT

Alesky, Botnik, and Clutz are about to play a tournament for the chess championship of Transylvania. Each is to play one game apiece with his two opponents. In the event of a three-way tie, they will start over. If there is a two-way tie for first place, the two will continue to play until one scores a win. On the average, the players' abilities are as follows: Alesky, playing Botnik, wins 10% and draws 90% of the games. Alesky, playing Clutz, wins 20% and draws 80% of the games. Botnik, playing Clutz, wins 55% and loses 45% of the games.

Alesky is obviously the best player. He will wind up 10 games ahead out of every 100 played with Botnik, and 20 games ahead out of every 100 played with Clutz. Botnik is intermediate; he

will wind up 10 games behind out of every 100 played with
Alesky, but 10 games ahead out of every 100 played with Clutz.
Clutz is the worst, since he will wind up 20 games behind out of
every 100 played with Alesky, and 10 games behind out of every
100 played with Botnik. Compare Alesky's and Clutz's chances
of winning the tournament.

80

THE LOCK

"I believe you are unfamiliar with the penal system in our
country," said the warden, as he led the new prisoner to his cell.
"We find that it improves prison morale for each prisoner to have
a chance to end his sentence at any time. In your case, we have
set up a combination lock on your cell door. There are ten dials,
on which you can set up any ten-digit number. If you set up the
right one, the cell door will unlock and you will be free to leave."

"I see," said the prisoner. "Then if I try every possible number,
I'm sure to hit the right one."

"True," said the warden, "but even if you were able to change
the numbers at the rate of one per second without rest, it would
still take you a hundred years to hit the right combination. However, you could try numbers at random and have a chance of
choosing the right one. Or, you could search for the clue which we
always provide."

"What sort of clue?"

"Well, it might be almost anything. For example, one of our
prisoners was put in an escape-proof cell and told that he would
be pardoned if he could break out. He was also given permission
to keep any plants he wished in his cell."

"What became of him?"

The warden chuckled. "After more than two years, he suddenly
realized that some words may have more than one meaning. He
requested a poison ivy plant. Soon after receiving it, he broke out
—in a rash. Naturally, he received his pardon."

The warden unlocked the cell and ushered the prisoner in.
"Your cell contains a desk calculator and writing implements.
Good luck."

The prisoner was left alone. He tried a few combinations on the lock without success. What could the clue be? A thought struck him. It seemed worth a try. He made a few calculations, and then set up a number on the lock. The cell door opened and the prisoner strolled out, after serving less than an hour of his sentence.

What number did he try?

81

TREES

How many trees can Farmer Ferd plant on his 100 foot square field if they are to be no closer than 10 feet apart? Neglect the thickness of the trees, and assume that trees may be planted on the boundary of the field.

82

REDUCING THE FRACTION

To keep the class occupied for several minutes, the teacher wrote a fraction on the blackboard and asked them to reduce it to its lowest terms. He followed a certain pattern in constructing the fraction, which permitted him to make it as long as he wished. He chose to stop after ten digits in the numerator and in the denominator. Neither numerator nor denominator ended in zero.

After a short time, while the rest of the class was still working on the calculations, Barney Boob was finished. The teacher looked at his paper and saw that he had the correct answer. However, the teacher was chagrined to find that Barney had arrived at the answer by the ridiculous procedure of "cancelling" any digit that appeared in both the numerator and the denominator. Furthermore, he would still have arrived at the correct answer if the teacher had chosen to continue with the system he was following and to stop after some other number of digits in both the numerator and denominator.

What was the fraction?

83

HAVING A BALL

Johnny's daddy has presented him with a solid white ball. Johnny has painted it black, but has now decided he would rather have a white ball, with a one-inch black stripe running around it. The ball is one foot in diameter. Show Daddy how he can cut the ball into four pieces which can be fitted together to form a white ball with a one-inch black stripe.

84

MORE BASEBALL

The Detroit Tigers and the New York Yankees have each played the same number of games so far this season. The Tigers have a .664 average, and the Yankees have won 70 games. Which team is ahead?

Note that each team plays 162 games a season, and that the team's average is the number of games won divided by the number of games played rounded off to three decimal places.

85

THE SOLAR SYSTEM

Which planet is usually closest to the planet Pluto? For purposes of analysis it may be assumed that the orbits are concentric circles in a plane, and that each planet has a different constant angular velocity. For a given position of any planet, each of the other planets has an equal probability of being at any point on its orbit.

86

HEXAGONAL NUMBERS

Hexagonal numbers are given by the relation $N = 3n^2 + 3n + 1$ where n is any positive integer. They may be formed by

n	0	1		2
N	1	7		19

placing *n* rows of points in the form of a hexagon about a central point. The first three hexagonal numbers, $N = 1$, 7, and 19, corresponding to $n = 0$, 1, and 2, respectively, are illustrated in the above sketch.

Barney Boob suggested the following even-money bet to Sammy Sharp: Barney would write a different digit on each of three pieces of paper and put them into a hat. Sammy would pull them out one at a time. If the three-digit number formed by writing the digits in the order in which they were drawn were a hexagonal number, Barney would win. Otherwise Sammy would win.

Sammy reasoned that there couldn't be many three-digit hexagonal numbers (there are, in fact, twelve), and that there are hundreds of different three-digit numbers altogether. Therefore, he gleefully accepted the bet.

Was this wise?

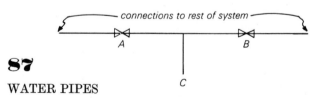

87
WATER PIPES

"If I open the cold water valve *A*, water comes out of the faucet *C* at 20°C," said Sam. "If instead I open the warm water valve B, water comes out of the faucet at 40°C. The trouble is, I need water at 60°C.''

"Well, 20 plus 40 is 60, so why not try opening both valves," suggested Barney.

Sam did so, and sure enough the water now came out of the faucet at 60°C. Explain how this could be.

Assume that the system consists of piping with constant flow resistances, and tanks of water at constant temperatures and pressures. The flow F and pressure drop ΔP across a given flow resistance R are related by $\Delta P = RF^2$. Neglect the variation in physical properties of water with temperature, and assume all water temperatures are in the range of 0°C to 100°C.

88

THE INVESTMENT COUNSELOR

"Boss, I have a complaint," said the young investment counselor. "Jones and I were hired at the same time, and both of us have handled about the same number of assignments. They've all been worth about the same, and each required a 'yes-or-no' decision. Now, I've been keeping track of our recommendations, and I've done pretty well—I've been right about 70% of the time. Jones though knows nothing about investments—he hasn't made the right recommendation more than 10% of the time. I know that you're as aware of this as I am, and yet you've given him a promotion and a raise, while turning me down. How come?''

Did the boss have any valid reason for his action?

89

FACTORS

Find two unequal numbers, A and B, such that $A + n$ is a factor of $B + n$ for all values of n from 0 through 10.

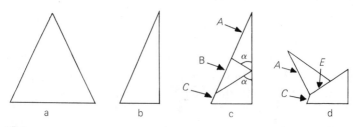

90

THE CONE

A cone (Fig. a) is cut in half axially (Fig. b). A wedge, B, making equal angles, α, with the axis as shown in Fig. c is cut out of it. The wedge is removed, and the upper piece, A, is placed on the lower piece, C, as shown in Fig. d. The question is, along the edge marked E in Fig. d, does the upper piece protrude past the lower piece, or does the lower one protrude past the upper?

91

THE PHILANTHROPIST

At the Millionaire's Club, Oliver Megabuck remarked that in the current charity campaign to help stamp out athlete's foot, he had pledged to match the total of all other contributions. Upon hearing this, Winston Lucre fainted. Can you guess why?

92

FIND THE NUMBER

n is a number between 0 and 10. For a certain number A, when the 21 digits in A and A^n are added, the sum is 1. (If the sum contains more than one digit, add the digits together, and repeat the process if necessary until a single digit results.) What is n?

93

APPLES

At harvest time, the orchards of Mr. MacIntosh, Mr. Jonathan, and Mr. Delicious had yielded 314,827 apples, 1,199,533 apples, and 683,786 apples, respectively. While lunching with Jonathan the following Sunday, MacIntosh mentioned the number of apples he would have left over if he divided his harvest equally among all the apple dealers.

"Why don't you sell those extra apples to me," suggested Jonathan, "and then I'll be able to divide my apples equally among all the dealers."

"Sorry," said MacIntosh, "but Mr. Delicious made the same suggestion for the same reason, and I've already accepted his offer."

How many apple dealers are there?

94

THE THREE-WAY DUEL

Anderson, Barnes, and Cramer are to fight a gun duel. They will stand close to one another so that each may kill one of the others or deliberately miss. The first to fire will be chosen at random, and they will rotate in the order Anderson, Barnes, Cramer, each firing one shot at a time. If there is more than one survivor after a number of rounds, one of the contenders will be chosen at random and required to shoot one of the others.

Before the duel starts, Anderson may make any statement, followed by a statement from Barnes, and finally one from Cramer. They will adhere to the following rules:

(1) A contender may not break any commitment he makes in his statement.

(2) He will act in his own best interest when it does not conflict with rule 1.

(3) He will act randomly when it does not conflict with rules 1 and 2.

There are referees to assure that the rules are adhered to. If a contender commits himself to a choice of actions on a statistical basis (for example, if Anderson commits himself to miss with a probability of 1/3), the choice will be made objectively (for example, by tossing dice).

What is Anderson's best strategy and his probability of surviving?

95

7 + 8 = 12

How can 7 + 8 = 12? One possibility is that the number system being used is to the base 13. However, what we are looking for here is a different digit to be substituted for each of the letters in the following example in order to give a correct addition.

SEVEN
<u>EIGHT</u>
<u>TWELVE</u>

96

PRESIDENTS

"Here is an odd item, Professor Flugel," said Tom, looking up from his newspaper. "It says here that three of the first five presidents of the United States died on the Fourth of July. I wonder what the odds are against a coincidence like that."

"I'm not sure," replied the professor, "but I'm willing to give ten–to–one odds I can name one of the three who died on that date."

Assuming that the professor had no prior knowledge of the dates on which any of the presidents died, was he justified in offering such odds?

97

THE THREE-WAY RACE

Boston, Detroit, and New York are tied for first place, with one game left in the season. In the last game, Boston is to play Detroit, and New York is to play Cleveland. If there is a tie for first place at the end of the regular season, there will be a single game play-off.

New York has a 55% probability of beating Cleveland, and a 55% probability of beating Boston or Detroit in a play-off. Detroit has a 45% probability of beating Boston, and a 45% probability of beating New York in a play-off. Compare New York's and Detroit's chances of winning the division championship.

98

CITIES

If the city of Aardvosk is 9000 miles from Baltimore, and Baltimore is 9000 miles from the city of Crupnik, what is the probability that Crupnik is closer to Aardvosk than to Baltimore?

99

STILL MORE TRUTH AND LIES

On one of his journeys Professor Flugel visited the country Aristotelia, where every inhabitant answers any question if he possibly can, but the answer may be either the truth or a lie, depending on his mood at the time.

Upon meeting any inhabitant, the professor was able to determine the fellow's name by asking him one question. What was it?

100

LUCKY THIRTEEN

This final problem is dedicated to those who find themselves unable to solve any puzzle, no matter how simple it appears.

Find a number, N, which has the following characteristics:

(a) N^{13} has the same final digit as N.

(b) The sum of the digits of N^{13} is equal to the sum of the digits of N^{31}. (If the sum contains more than one digit, add the digits together, and repeat the process if necessary until a single digit results.)

(c) $N^{13} - N$ is divisible by 13.

SOLUTIONS

1. POKER HANDS

Hand (d) is the best, hand (a) is the worst, and hands (b) and (e) are equal in strength.

If the hands could occur during the same deal, then of course (a) would be best, (b) and (c) would be equal, and (e) would be worst. However, as the hands cannot occur on the same deal, their values must be based on their probabilities of winning. Each of the full houses can be beaten by the same number of four-of-a-kinds, but by different numbers of straight flushes. For example, hand (a) can be beaten by 7 straight flushes in spades, 7 in hearts, 8 in diamonds, and 10 in clubs, for a total of 32; while hand (d) can be beaten by 3 in spades, 8 in hearts, 8 in diamonds, and 5 in clubs, for a total of 24. A complete tabulation follows:

AS AH AD PLUS		STRAIGHT FLUSHES THAT BEAT IT				
		S	H	D	C	Total
KS KH,	or 2S 2H	7	7	8	10	32
KS KC,	or 2S 2C	7	8	8	8	31
QS QH,	or 3S 3H	6	6	8	10	30
QS QC,	or 3S 3C	6	8	8	7	29
JS JH,	or 4S 4H	5	5	8	10	28
JS JC,	or 4S 4C	5	8	8	6	27
10S 10H,	or 5S 5H	4	4	8	10	26
10S 10C,	or 5S 5C	4	8	8	5	25
9S 9H,	or 6S 6H	3	3	8	10	24
9S 9C,	or 6S 6C	3	8	8	5	24
8S 8H,	or 7S 7H	3	3	8	10	24
8S 8C,	or 7S 7C	3	8	8	5	24

2. 100 DIGITS

The probability is zero. If a number has n digits, then its square has either $2n$ or $2n-1$ digits. Therefore, a number plus its square has either $3n$ or $3n-1$ digits. Since 100 is of the form $3n+1$, it is impossible to form a number and its square using exactly 100 digits.

3. THE CHESS MATCH

Abner has a better than 53% probability of winning the match. His best strategy is to play daringly if he is behind or tied, and conservatively if he is ahead.

The possible outcomes and their probabilities are sketched below. W, D, and L represent Abner's wins, draws and losses.

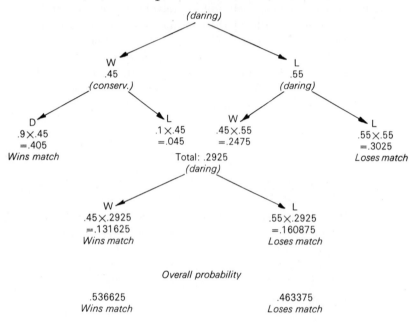

4. THE PROFESSOR'S TELEPHONE NUMBER

The professor's telephone number is 9899901, which is equal to 9×1099989. Other numbers which have this characteristic may be formed by inserting any number of nines between the 98 and the 01, i.e. 9801, 98901, 989901, etc. Another number of this type is 8712 which is equal to 4×2178. Finally, additional numbers may be formed by repeating the basic groups, e.g.,

$$98999019801989901 = 9 \times 10998910891099989.$$

5. HOUSES AND WELLS

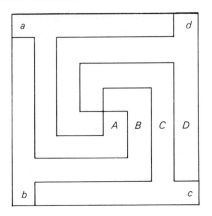

6. FOUR POINTS

There is nothing in the wording of the problem which restricts the points to a plane. Billy placed the four points at the corners of a tetrahedron.

7. THE CASE OF THE CURIOUS CUBE

The face opposite the "six" must show two spots. An expanded view of the cube is shown below.

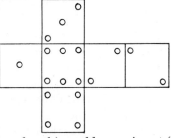

Most people attempting to solve this problem arrive at "one" as the solution. The reason that this is incorrect is that the "two" and "three" shown in view 1 cannot both be the same faces as are shown in view 2. If the cube is held so that the "two" and "three" form a V, then the "two" is on the left in view 1, and on the right in view 2.

8. HEXAGONS

Once it is realized that the hexagon need not be convex, it is easy to find any number of solutions similar to the one shown below.

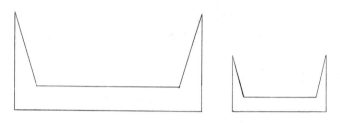

9. AN UNUSUAL NUMBER

A number which is a square, cube, and fifth power must also be a $2 \times 3 \times 5 = 30$th power. 3^{30} is somewhat greater than two hundred trillion, so that the solution must be 2^{30} which is equal to 1,073,741,824.

10. TOSSING COINS

(a) Boris (16)
(b) Charlie (20)
(c) Abner (18)

For an even number of tosses, the possibility of the same number of heads and tails being tossed reduces the probability of one being tossed more than the other. The higher the number of even tosses, the lower the probability of the same number of heads and tails. Therefore, the one least likely to have tossed more heads than tails is the one with the lowest number of even tosses.

11. THE CHESS GAME

Since chess is played with a white square at each player's near right corner of the board, the players must be at the left and right sides of the board illustrated. Therefore, whether white is moving to the left or to the right, he wins by queening his pawn: P–B8 (Queen), mate.

12. FATHER AND SON

Hubert is three and his father is 51. $3 \times 51 = 153$.

13. THE GAMBLER

Whenever he bets $10 on red, she bets $100 on black.

14. JOHN'S COUSINS

John is 43 and his cousins are 2, 2, and 3. For any other value of John's age, more than one combination would have existed, and the professor would not have had sufficient information to determine the cousins' ages.

MULTIPLICATIONS

15
$$\begin{array}{r} 927 \\ 63 \\ \hline 58401 \end{array}$$

16
$$\begin{array}{r} 402 \\ 39 \\ \hline 15678 \end{array}$$

17
$$\begin{array}{r} 715 \\ 46 \\ \hline 32890 \end{array}$$

18
$$\begin{array}{r} 297 \\ 54 \\ \hline 16038 \end{array}$$

19
$$\begin{array}{r} 594 \\ 27 \\ \hline 16038 \end{array}$$

20
$$\begin{array}{r} 396 \\ 45 \\ \hline 17820 \end{array}$$

21
$$\begin{array}{r} 345 \\ 78 \\ \hline 26910 \end{array}$$

22
$$\begin{array}{r} 367 \\ 52 \\ \hline 19084 \end{array}$$

23
$$\begin{array}{r} 495 \\ 36 \\ \hline 17820 \end{array}$$

24. DISSECTING THE CHESSBOARD

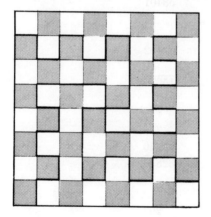

25. FOUL PLAY

Detroit, Boston, and a third team, New York, are tied for first place. Boston's last game is with New York, while Detroit's last game is with one of the remaining teams. If Detroit loses and Boston wins, then Boston has won the division championship; while if Detroit wins, then it winds up in a tie with the winner of the Boston–New York game.

26. WEATHER REPORT

The five temperatures were 1, -1, 2, -2, and 3.

27. WHAT PRICE LUNCH?

In their state, sales tax is four cents on purchases of 76¢ to $1.10, and five cents on purchases of $1.11 to $1.35. Therefore, a bill of $1.15 cannot occur.

28. DIGITS EQUAL POWER

$9^{21} = 109{,}418{,}989{,}131{,}512{,}359{,}209.$

For values of n larger than 21, 10^n has $n + 1$ digits, while 9^n has less than n digits.

29. ADDITIONS

The two solutions are

$$
\begin{array}{r@{\qquad\qquad}r}
146 & 157 \\
583 & 482 \\
\hline
729 & 639 \\
\end{array}
$$

and

$$
\begin{array}{r@{\qquad\qquad}r}
718 & 729 \\
236 & 135 \\
\hline
954 & 864 \\
\end{array}
$$

30. ADDITIONAL ADDITIONS

583	715
146	248
729	963

31. SQUARE PATTERNS

Sammy's reasoning has a fallacy. The squares do not appear the same number of times in the 16 2 × 2 arrays. Each corner square appears once, the edge squares appear twice, while interior squares appear four times. Therefore, the color which appears fewer times can compensate by having greater representation in the interior. Two solutions are shown below.

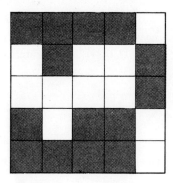 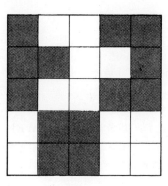

32. A BAG OF MARBLES

There is a probability of 2/3 that the second marble would also be purple. This result is independent of the number of marbles in the bag. The proof is as follows:

Let n = number of marbles in bag.

Each of the following $n + 1$ combinations has a probability of $1/(n + 1)$.

No. of oranges	No. of purples
0	n
1	$n - 1$
2	$n - 2$

$$
\begin{array}{cc}
\cdots & \cdots \\
n - k & k \\
\cdots & \cdots \\
n & 0
\end{array}
$$

There are a total of $n(n + 1)/2$ purple marbles in all of the above combinations, each of which has an equal probability, $2/n(n + 1)$, of being picked first, if the first one picked is specified as being purple. Thus, the probability that the first marble came from the bag with a given number, k, of purples is $2k/n(n + 1)$. After the first one is picked, there are $k - 1$ purples among the $n - 1$ marbles left in the bag, and the probability of the second pick being purple is $(k - 1)/(n - 1)$. The probability that the first purple marble came from that bag and that the second marble was also purple is the product of the individual probabilities, $2k(k - 1)/n(n + 1)(n - 1)$. The overall probability for all of the possible combinations is

$$
\sum_{k=0}^{n} \frac{2k(k - 1)}{n(n + 1)(n - 1)} = \frac{2(1/3)(n - 1)n(n + 1)}{n(n + 1)(n - 1)} = \frac{2}{3}
$$

33. ELLIPSE

The largest (and only) ellipse is the border of the pond which encloses an area of fifty square feet. The other two ovals are not ellipses.

34. ANOTHER CURIOUS CUBE

The bottom face in Figure 1, opposite the face colored A, must also be colored A.

35. THE RACE FOR THE DIVISION CHAMPIONSHIP

The probability of Detroit winning the championship is about 51%. The proof of this curious result is as follows: The probability of Detroit winning all three games to win the championship is $.66^3 = .287496$. The probability of Detroit winning two and losing one to tie for the pennant is $3 \times .66^2 \times .34 = .444312$. The remainder, $.268192$, is the probability of Detroit losing two or three games to lose the championship. The case of the tie results in equal probabilities of .222156 of Detroit winning or losing the play-off. Therefore, the overall probability of Detroit winning the championship is .509652.

36. A GROUP OF NUMBERS

There cannot be more than 50 integers in the group or there will be some pair totaling 100. There are a number of solutions of 50 integers, of which the simplest is the group of integers 50 through 99.

37. TWO ADDITIONS

286	286
286	3210
572	3496

38. REGARDING TRUTH AND LIES

The professor's question was, "If I asked you, 'Does a triangle have four sides?' would your answer be the same as your answer to this question?" Neither a person who always lies nor one who is always truthful could answer "Yes" or "No" without breaking his rule. Any question for which the true answer is "No" could of course be substituted for "Does a triangle have four sides?" The professor could also have asked, "If I asked you, 'Does a triangle have three sides?' would your answer be different from your answer to this question?" and achieved the same result.

39. FIVE WEIGHTS

Call the weights A, B, C, D and E.
The first three weighings are

    1. AB vs CD
    2. AC vs BD
    3. AD vs BC

There are three possible situations:

I. None of the first three weighings balance, and there is one weight (e.g., A) which is always on the heavier side; i.e.,

Therefore, $A = 5$ and $E = 4$.

The next two weighings are

4. B vs C
5. E vs BC

Assume that, say, B is found to be heavier than C in the fourth weighing. Then from the result of the final weighing

$B =$	2	3	3
$C =$	1	1	2
$D =$	3	2	1

II. None of the first three weighings balance, and there is one weight (e.g., A) which is always on the lighter side; i.e.,

1. $AB \searrow CD$ 2. $AC \searrow BD$ 3. $AD \searrow BC$

Therefore, $A = 1$ and $E = 2$.
The next weighing is

4. B vs C

Assume that, say, B is found to be heavier than C. The final weighing is

5. B vs AD

and the result is

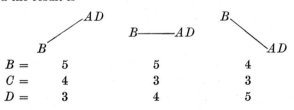

$B =$	5	5	4
$C =$	4	3	3
$D =$	3	4	5

III. One of the weighings (e.g., weighing No. 1) balances; in the two remaining weighings, one of the weights (e.g., A) is on the heavy side, and one of the weights (e.g., B) is on the light side; i.e.,

1. AB———CD 2. AC⟋BD 3. AD⟋BC

The next weighing is

4. C vs D

Assume that, say, C is found to be heavier than D. The final weighing is

5. E vs BD

and the result is

	E⟋BD	E———BD	E⟍BD
$A =$	4	5	5
$B =$	1	1	2
$C =$	3	4	4
$D =$	2	2	3
$E =$	5	3	1

40. FEWEST WEIGHINGS

There are $5! = 120$ possible permutations of the five weights. Each weighing separates the possibilities into one of three categories—left side heavier, right side heavier, or both sides balance. Even with the most efficient system of sorting the possibilities into three categories, no more than 3^n initial possibilities can be reduced to a single possibility in n weighings. Since $3^4 = 81$ and $3^5 = 243$, it requires at least five weighings to determine which of the initial 120 possibilities is correct.

MORE ADDITIONS

41

$$\begin{array}{r} 347 \\ 859 \\ \hline 1206 \end{array}$$

42	437
	589
	1026

43	426
	879
	1305

44	246
	789
	1035

45	624
	879
	1503

46	264
	789
	1053

47	743
	859
	1602

48	473
	589
	1062

49

4
35
987
――――
1026

50

3
45
978
――――
1026

51

3
74
985
――――
1062

52. ANOTHER CHESS GAME

The key to this problem lies in considering the moves that could have led to the position shown. Black could not have just moved his king. If the king had moved from any other square, it would already have been in check prior to white's previous move. He could not have moved his bishop or the three pawns on their starting squares. Black's pawn on B4 could not have moved from B3 because white would have been in check, and could not have arrived by capturing a piece. The only possible move black could have made is P(B2)–B4. Therefore, white wins by P × P (en passant), mate.

53. THE PERFECT SQUARE

The last digit or pair of digits of a perfect square must be 1, 4, 6, 9, 25, or 00. The sum of all the digits must be 1, 4, 7, or 9. (If the sum contains more than one digit, add the digits together, and repeat the process if necessary until a single digit results.)

This eliminates the first four numbers, and so the perfect square must have been the completely missing fifth number.

54. OBJECT IN MOTION

The object is in orbit. Friction with the atmosphere causes it to drop to a lower orbit where the orbital speed is higher. The decrease in potential energy compensates for the increase in speed as well as the energy lost as friction.

55. ANGLES

Archie made angle B first, and constructed angle A three times as large.

56. THE BETTER HAND

As in Problem 1, the hands cannot occur during the same deal, and therefore their values must be based on their probabilities of winning. Each full house can be beaten by the same number of four-of-a-kinds, but by a different number of straight flushes. Hand (a) can be beaten by 32 straight flushes (7 in spades, 7 in hearts, 8 in diamonds, and 10 in clubs), while hand (b) can be beaten by only 31 straight flushes (7 in spades, 8 in hearts, 8 in diamonds, and 8 in clubs). Therefore, hand (b) is better.

57. PYTHAGOREAN PARALLELEPIPED

In the two examples of Pythagorean triangles which were given, $3^2 + 4^2 = 5^2$ and $5^2 + 12^2 = 13^2$, substitute the first in the second to obtain $3^2 + 4^2 + 12^2 = 13^2$.

58. THE TWO-TONE CUBE

Developed views of the two cubes are shown below. No matter which face is uppermost, the pattern formed by the others is the same.

59. THE BASEBALL SEASON

Because of postponements, Washington has enough games left to play so that it can tie for the division championship if it wins them all, and would have a chance to beat New York in a play-off. Detroit, although ahead of Washington in percentages, has already been eliminated. For example, the standings may be as follows:

	WINS	LOSSES	%	GAMES BEHIND	GAMES LEFT TO PLAY
New York	108	54	.6667	—	0
Detroit	106	55	.6584	$1\frac{1}{2}$	1
Washington	104	54	.6582	2	4

60. ROULETTE

The catch is that for cases where the number of wins and losses do not match, then the amount of money that would be lost in the event of an appreciable excess number of losses would be greater than the amount which would be won in the event of an equivalent excess number of wins. If all possible outcomes of a given number of plays are considered, the average results in the house collecting its usual percentage. The "system" could have used percentages other than 20%. If 100% had been used, it would have been the well-known system of doubling your bet whenever you lose.

61. THE NUMBER RING

The four solutions are:

(a)	1	3	5	7	9	11	2	4	6	8	10
(b)	1	4	7	10	2	5	8	11	3	6	9
(c)	1	5	9	2	6	10	3	7	11	4	8
(d)	1	6	11	5	10	4	9	3	8	2	7

It is interesting to note that any pair of the four solutions also satisfy the conditions of the problem.

62. THE POKER GAME

The professor replied that Jim's hand with four-of-a-kind was the worst. Each of his four opponents will lose to John's royal flush. Therefore, the worst hand is the one which is likely to lose the most money. Joe, Jason, and Jerry, with weak hands, are more likely to drop out early in the betting, so that Jim is most likely to be the biggest loser.

MORE MULTIPLICATIONS

63 $1 \times 26 \times 345 = 8970$

64 $\qquad 2 \times 14 \times 307 = 8596$

65. AREA OF ANNULUS

The center of the circle and points A and B form a right triangle. Let r_1 = radius of inner circle and r_2 = radius of outer circle. From the Pythagorean Theorem, $r_2^2 - r_1^2 = 1$. The area of the annular region is $\pi(r_2^2 - r_1^2)$. Since the term in parentheses equals unity, the area of the annular region is π square inches.

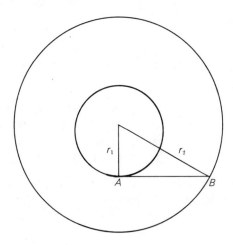

66. MORE TRUTH AND LIES

The question is:

"Are both of the following statements true?:

(a) You are type A.

(b) If this question were, 'Does a triangle have four sides?' your answer would be the same as your present answer."

In response to this question, type T will answer, "No"; type F will answer, "Yes"; and type A will be unable to reply.

67. ROWS OF BUSHES

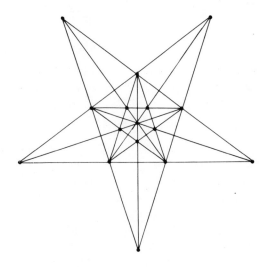

68. THE TENTH POWER

The simplest approach is to note that only two ten-digit numbers are tenth powers: 8^{10} and 9^{10}. Furthermore, by "casting out nines" it is found that the ten-digit number is a multiple of 9, so that only 9^{10} need be considered. When 9^{10} is evaluated, the result is found to contain several repeated digits, so that the assumption that no solution exists is verified.

69. CARD TRICK

(a) There were thirty cards in the original pack.

(b) When the trick was repeated with 32 cards, the final order of the cards was the reverse of the original order.

STILL MORE MULTIPLICATIONS

70
$$\begin{array}{r} 297 \\ 18 \\ \hline 5346 \end{array}$$

71
$$\begin{array}{r} 483 \\ 12 \\ \hline 5796 \end{array}$$

72
$$\begin{array}{r} 138 \\ 42 \\ \hline 5796 \end{array}$$

73
$$\begin{array}{r} 157 \\ 28 \\ \hline 4396 \end{array}$$

74
$$\begin{array}{r} 186 \\ 39 \\ \hline 7254 \end{array}$$

75
$$\begin{array}{r} 159 \\ 48 \\ \hline 7632 \end{array}$$

76
$$\begin{array}{r} 198 \\ 27 \\ \hline 5346 \end{array}$$

77. THE DEPARTMENT STORE

While she was waiting for him to find her in the hat department, he was following the same reasoning and waiting for her to find him in the book department.

78. HALF A CHESSBOARD

The original order of the rows was (d), (a), (b), (c), and the number originally in the third row was 26382629.

For numbers which are factorable by 11, alternately adding and subtracting the digits also results in a number factorable by 11. The colors of the squares identify the alternate digits, from which (c) is seen to be the only number factorable by 11, no matter what the order. Numbers (a) and (c) are the only ones containing a 5 or a 0. Since (c) is already identified as the number factorable by 11, (a) must be the number factorable by 5. From the sum of the digits, (a) and (d) are factorable by 3. Since (a) is already identified as the number factorable by 5, (d) must be the top number. Thus the order of the numbers are (d), (a), (b), (c). Since (a) ends in a black square, 0, then (b), the number factorable by 7, must end in a white square. There are 48 possible permutations, but the only one factorable by 7 is 26382629.

79. THE CHESS TOURNAMENT

Clutz has a 32.4% probability of winning the tournament, compared with Alesky's 28% probability of winning. This paradoxical result is proved by the following tabulation:

| OUTCOME | | | SCORE | | | PROBABILITY |
| A–B | A–C | B–C | Winner in bold face | | | |
			A	B	C	
1 0	1 0	1 0	**2**	1	0	$(.1)(.2)(.55) = .011$
1 0	1 0	0 1	**2**	0	1	$(.1)(.2)(.45) = .009$
1 0	½ ½	1 0	$1\frac{1}{2}$	1	½	$(.1)(.8)(.55) = .044$
1 0	½ ½	0 1	$1\frac{1}{2}$*	0	$1\frac{1}{2}$	$(.1)(.8)(.45) = .036$
½ ½	1 0	1 0	$1\frac{1}{2}$*	$1\frac{1}{2}$	0	$(.9)(.2)(.55) = .099$
½ ½	1 0	0 1	$1\frac{1}{2}$	½	1	$(.9)(.2)(.45) = .081$
½ ½	½ ½	1 0	1	$1\frac{1}{2}$	½	$(.9)(.8)(.55) = .396$
½ ½	½ ½	0 1	1	½	$1\frac{1}{2}$	$(.9)(.8)(.45) = .324$

(* means *wins play-off*)

80. THE LOCK

The clue was that the warden's statement was literally true. At the rate of one number per second, it would take precisely one hundred years to reach the right combination. The prisoner calculated that there are 3155760000 seconds in a hundred years and found that it worked.

81. TREES

The following arrangement shows how, by a combination of square and equilateral triangular spacings, the farmer can plant 128 trees in his field.

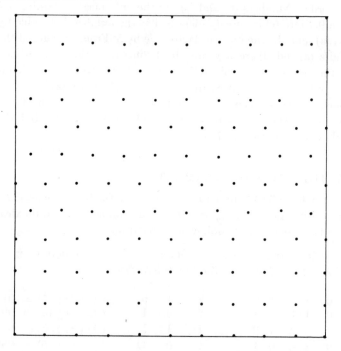

82. REDUCING THE FRACTION

$$\frac{1666666666}{6666666664} = \frac{1}{4}$$

The result is unchanged as long as the strings of 6's in the numerator and denominator have the same number of digits.

83. HAVING A BALL

The solution is shown below. Cross-sections through the center of the sphere are shown, and the cuts are symmetrical about the vertical axis of the sphere.

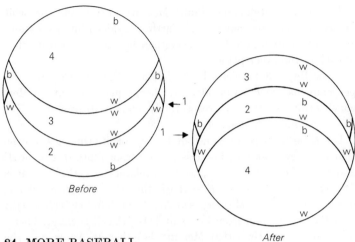

Before

After

84. MORE BASEBALL

A team has to play at least 107 games before it can have a
.664 average. For a smaller number of games, the average will be
greater or less than .664, but never exactly .664 to three decimal
places. For 107 or more games and a .664 average, the team must
have 71 or more wins. Therefore the Tigers lead the Yankees.

85. THE SOLAR SYSTEM

The planet which is usually closest to Pluto is Mercury. Note
that the question was not "Which planet approaches closest to
Pluto?" or "Which planet's orbit is closest to that of Pluto?"

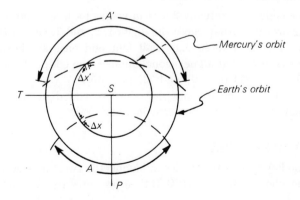

The above sketch shows Pluto, Mercury's orbit, and the orbit of any intermediate planet, e.g., Earth. Consider Pluto as stationary. Pluto's angular velocity may then be subtracted from that of each of the other planets.

Draw line PS between Pluto (P) and the sun (S), and TS perpendicular to PS. Consider two short equal segments of Mercury's orbit, Δx and $\Delta x'$, respectively, equally distant from TS. Arc A' is longer than arc A.

Let E be the circumference of Earth's orbit. During the time that Mercury is at $\Delta x'$, Earth is further from Pluto if it is on A' and closer if it is on $E - A'$; i.e., the probability is A'/E that it is further from Pluto and $(E - A')/E$ that it is closer. Similarly, during the time that Mercury is at Δx, the probability is $(E - A)/E$ that Earth is further from Pluto, and A/E that it is closer. Therefore, during the time that Mercury is either at Δx or $\Delta x'$, the probability is $(A' + E - A)/2E$ that Earth is further from Pluto, and $(E - A' + A)/2E$ that it is closer. Since A' is longer than A, there is a greater probability that Earth is further than Mercury from Pluto than that it is closer during the time that Mercury is within the segments Δx and $\Delta x'$.

Since Mercury's orbit may be completely subdivided into pairs of equal segments such as Δx and $\Delta x'$, and since any of the other planets may be substituted for Earth in the above discussion, it is concluded that Mercury is closer to Pluto more of the time than is any of the other planets.

86. HEXAGONAL NUMBERS

The odds were actually 2 to 1 in favor of a hexagonal number being drawn. Barney chose the digits 1, 2, and 7. Out of the six permutations of these digits, all four odd ones are hexagonal numbers (note that all hexagonal numbers are odd): 127 ($n = 6$), 217 ($n = 8$), 271 ($n = 9$), and 721 ($n = 15$).

It is interesting to note that the two odd digits chosen by Barney, 1 and 7, are also hexagonal numbers.

87. WATER PIPES

The key to this puzzle lies in having the two tanks K and L in the following sketch at 0°C. If their pressures are such as to give

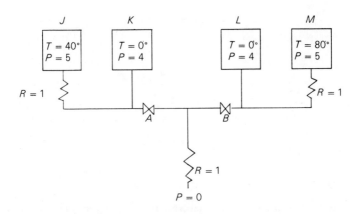

zero flow from these tanks when both valves are open, then for any combination of temperatures, pressures and resistances in the rest of the system, $T_A + T_B = T_{A,B}$. The subscripts indicate which valves are open. The values given in the above sketch are one example of how the desired solution can be achieved. The resulting flows and outlet temperatures are shown below.

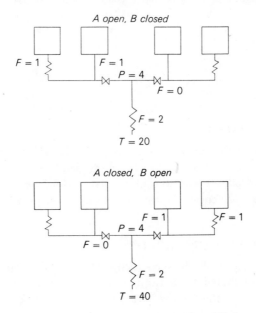

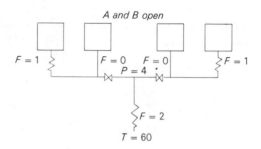

88. THE INVESTMENT COUNSELOR

Jones was one of their most valuable employees. They simply did the opposite of what he recommended and made handsome profits.

89. FACTORS

One solution is $A = 1$, $B = 11! + 1$.

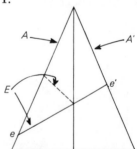

90. THE CONE

If one of the cuts had been made through the complete cone, the piece labeled A' would have been identical with A. The cut forms an ellipse with the major axis extending from e to e'. Therefore, the edges E of the upper and lower pieces coincide.

91. THE PHILANTHROPIST

Lucre had made the same pledge. As a result, even if one cent were contributed by a third individual, Megabuck and Lucre could donate their entire fortunes trying to match the other's contribution plus one cent.

92. FIND THE NUMBER

A and A^n cannot have a total of $m(n + 1) + 1$ digits, where m is any integer. Therefore, for 21 digits, n cannot be 1, 3 4, or 9. The values of n for which the sum of the digits can be 1 are 1, 3, 7, or 9. Since 1, 3, and 9 have been eliminated, n is 7.

93. APPLES

The number of dealers must be a divisor of $314,827 + 1,199,533 = 1,514,360$, and of $314,827 + 683,786 = 998,613$. The greatest common divisor of these two numbers is 131. Since 131 is a prime, there are 131 dealers.

94. THE THREE-WAY DUEL

By adopting the following strategy, Anderson's probability of surviving approaches certainty!

Anderson makes the following statement: "If Barnes agrees to shoot Cramer, then I will do the following:

(a) If I shoot first, then I will miss.

(b) If Barnes shoots first, then I will shoot Barnes with a probability of .9999 and will miss with a probability of .0001.

(c) If Cramer shoots first and shoots Barnes, then I will shoot Cramer with a probability of .9999 and will miss with a probability of .0001. If Cramer misses, then I will miss.

"If Barnes states anything other than that he will shoot Cramer, then if Cramer agrees to shoot Barnes, I will shoot Barnes, or will fire to miss if Barnes has already been shot. If Cramer does not agree to shoot Barnes, then I will shoot Cramer."

As a result of Anderson's statement, Barnes will agree to shoot Cramer, and Cramer, if he shoots first, will shoot Barnes. Barnes and Cramer will each have a .00003 probability of surviving and Anderson will have a .99993 probability of surviving.

Barnes will agree to shoot Cramer because otherwise he will have no chance of surviving. For if Barnes does not agree to shoot Cramer, then Cramer will have at least a 2/3 probability of surviving by agreeing to shoot Barnes, and Cramer will have less than a

2/3 probability of surviving by making any deal which gives Barnes any probability of surviving.

If Cramer shoots first he will shoot Barnes, as otherwise he will have no chance of surviving.

The results of this problem are quite remarkable in that an arrangement is readily conceivable in which, for example, Barnes and Cramer would each have a .5 probability of surviving and Anderson would have no chance of surviving. However, as much as Barnes and Cramer would desire to make such an arrangement, there is no way for them to do so.

95. $7 + 8 = 12$

$$\begin{array}{r} 69298 \\ 90431 \\ \hline 159729 \end{array}$$

96. PRESIDENTS

If the fifth president were not among those who died on that date, then the newspaper item would almost certainly have made the more impressive statement that, "Three of the first four presidents died on the Fourth of July." Therefore, Professor Flugel was reasonably confident that the fifth president, James Monroe, died on that date.

97. THE THREE-WAY RACE

Detroit has about a 31.39% probability of winning the championship as compared with New York's 30.25%. The reason is that New York has to win two games to win the championship, while Detroit can win in one game if New York loses its first game. This more than compensates for New York's better chances in each individual game.

98. CITIES

The probability is 100%. Taking the circumference of the Earth as 25,000 miles, Crupnik is anywhere from zero to 7000 miles from Aardvosk.

99. STILL MORE TRUTH AND LIES

Professor Flugel asked the following question: "If I asked you, 'What is your name?' and the truth or falseness of your answer were the same as for your answer to the present question, what is a possible answer you might give?"

The only way the inhabitant could tell the truth or could lie in his reply would be by giving his name; any other reply would be neither the truth nor a lie.

For example, suppose he decided to lie. If he were asked, "What is your name?" and also decided to lie, then he could give any answer other than his name. Therefore, in order to lie in answer to Professor Flugel's question, the only answer he could give would be his name.

On the other hand, if he decided to tell the truth, then his answer would be the same as a truthful answer to the question, "What is your name?" which would also be his name.

100. LUCKY THIRTEEN

Every number has these characteristics.

(a)

IF THE FINAL DIGIT OF N IS	THEN N^m WILL HAVE THE SAME FINAL DIGIT IF m IS OF THE FORM
0	n (i.e., any value)
1	n
2	$4n + 1$
3	$4n + 1$
4	$2n + 1$
5	n
6	n
7	$4n + 1$
8	$4n + 1$
9	$2n + 1$

13 is of each of the forms n, $2n + 1$, and $4n + 1$, and therefore, N^{13} always has the same final digit as N.

(b)

If the sum of the digits of N is	Then the sum of the digits of N^m will be the same if m is of the form
1	n
2	$6n + 1$
4	$3n + 1$
5	$6n + 1$
7	$3n + 1$
8	$2n + 1$
9	n

The sum of the digits will be 9 if the sum of the digits of N is 3 or 6 and m is greater than 1. Since 13 and 31 are both of each of the forms n, $2n + 1$, $3n + 1$, and $6n + 1$, N^{13} and N^{31} will both have the same sum of digits as N, unless the sum of digits of N is 3 or 6, in which case the sum of digits of N^{13} and N^{31} will be 9.

(c) Fermat's theorem (not to be confused with "Fermat's last theorem"—see any text on the theory of numbers) states that $N^p - N$ is divisible by p if p is a prime. Therefore, since 13 is a prime, $N^{13} - N$ is divisible by 13 for any value of N.

A CATALOGUE OF SELECTED DOVER BOOKS
IN ALL FIELDS OF INTEREST

AMERICA'S OLD MASTERS, James T. Flexner. Four men emerged unexpectedly from provincial 18th century America to leadership in European art: Benjamin West, J. S. Copley, C. R. Peale, Gilbert Stuart. Brilliant coverage of lives and contributions. Revised, 1967 edition. 69 plates. 365pp. of text.

21806-6 Paperbound $3.00

FIRST FLOWERS OF OUR WILDERNESS: AMERICAN PAINTING, THE COLONIAL PERIOD, James T. Flexner. Painters, and regional painting traditions from earliest Colonial times up to the emergence of Copley, West and Peale Sr., Foster, Gustavus Hesselius, Feke, John Smibert and many anonymous painters in the primitive manner. Engaging presentation, with 162 illustrations. xxii + 368pp.

22180-6 Paperbound $3.50

THE LIGHT OF DISTANT SKIES: AMERICAN PAINTING, 1760-1835, James T. Flexner. The great generation of early American painters goes to Europe to learn and to teach: West, Copley, Gilbert Stuart and others. Allston, Trumbull, Morse; also contemporary American painters—primitives, derivatives, academics—who remained in America. 102 illustrations. xiii + 306pp. 22179-2 Paperbound $3.00

A HISTORY OF THE RISE AND PROGRESS OF THE ARTS OF DESIGN IN THE UNITED STATES, William Dunlap. Much the richest mine of information on early American painters, sculptors, architects, engravers, miniaturists, etc. The only source of information for scores of artists, the major primary source for many others. Unabridged reprint of rare original 1834 edition, with new introduction by James T. Flexner, and 394 new illustrations. Edited by Rita Weiss. 6⅝ x 9⅝.

21695-0, 21696-9, 21697-7 Three volumes, Paperbound $13.50

EPOCHS OF CHINESE AND JAPANESE ART, Ernest F. Fenollosa. From primitive Chinese art to the 20th century, thorough history, explanation of every important art period and form, including Japanese woodcuts; main stress on China and Japan, but Tibet, Korea also included. Still unexcelled for its detailed, rich coverage of cultural background, aesthetic elements, diffusion studies, particularly of the historical period. 2nd, 1913 edition. 242 illustrations. lii + 439pp. of text.

20364-6, 20365-4 Two volumes, Paperbound $6.00

THE GENTLE ART OF MAKING ENEMIES, James A. M. Whistler. Greatest wit of his day deflates Oscar Wilde, Ruskin, Swinburne; strikes back at inane critics, exhibitions, art journalism; aesthetics of impressionist revolution in most striking form. Highly readable classic by great painter. Reproduction of edition designed by Whistler. Introduction by Alfred Werner. xxxvi + 334pp.

21875-9 Paperbound $2.50

DESIGN BY ACCIDENT; A BOOK OF "ACCIDENTAL EFFECTS" FOR ARTISTS AND DESIGNERS, James F. O'Brien. Create your own unique, striking, imaginative effects by "controlled accident" interaction of materials: paints and lacquers, oil and water based paints, splatter, crackling materials, shatter, similar items. Everything you do will be different; first book on this limitless art, so useful to both fine artist and commercial artist. Full instructions. 192 plates showing "accidents," 8 in color. viii + 215pp. 8⅜ x 11¼. 21942-9 Paperbound $3.50

THE BOOK OF SIGNS, Rudolf Koch. Famed German type designer draws 493 beautiful symbols: religious, mystical, alchemical, imperial, property marks, runes, etc. Remarkable fusion of traditional and modern. Good for suggestions of timelessness, smartness, modernity. Text. vi + 104pp. 6⅛ x 9¼. 20162-7 Paperbound $1.25

HISTORY OF INDIAN AND INDONESIAN ART, Ananda K. Coomaraswamy. An unabridged republication of one of the finest books by a great scholar in Eastern art. Rich in descriptive material, history, social backgrounds; Sunga reliefs, Rajput paintings, Gupta temples, Burmese frescoes, textiles, jewelry, sculpture, etc. 400 photos. viii + 423pp. 6⅜ x 9¾. 21436-2 Paperbound $4.00

PRIMITIVE ART, Franz Boas. America's foremost anthropologist surveys textiles, ceramics, woodcarving, basketry, metalwork, etc.; patterns, technology, creation of symbols, style origins. All areas of world, but very full on Northwest Coast Indians. More than 350 illustrations of baskets, boxes, totem poles, weapons, etc. 378 pp. 20025-6 Paperbound $3.00

THE GENTLEMAN AND CABINET MAKER'S DIRECTOR, Thomas Chippendale. Full reprint (third edition, 1762) of most influential furniture book of all time, by master cabinetmaker. 200 plates, illustrating chairs, sofas, mirrors, tables, cabinets, plus 24 photographs of surviving pieces. Biographical introduction by N. Bienenstock. vi + 249pp. 9⅞ x 12¾. 21601-2 Paperbound $4.00

AMERICAN ANTIQUE FURNITURE, Edgar G. Miller, Jr. The basic coverage of all American furniture before 1840. Individual chapters cover type of furniture—clocks, tables, sideboards, etc.—chronologically, with inexhaustible wealth of data. More than 2100 photographs, all identified, commented on. Essential to all early American collectors. Introduction by H. E. Keyes. vi + 1106pp. 7⅞ x 10¾. 21599-7, 21600-4 Two volumes, Paperbound $11.00

PENNSYLVANIA DUTCH AMERICAN FOLK ART, Henry J. Kauffman. 279 photos, 28 drawings of tulipware, Fraktur script, painted tinware, toys, flowered furniture, quilts, samplers, hex signs, house interiors, etc. Full descriptive text. Excellent for tourist, rewarding for designer, collector. Map. 146pp. 7⅞ x 10¾. 21205-X Paperbound $2.50

EARLY NEW ENGLAND GRAVESTONE RUBBINGS, Edmund V. Gillon, Jr. 43 photographs, 226 carefully reproduced rubbings show heavily symbolic, sometimes macabre early gravestones, up to early 19th century. Remarkable early American primitive art, occasionally strikingly beautiful; always powerful. Text. xxvi + 207pp. 8⅜ x 11¼. 21380-3 Paperbound $3.50

PLANETS, STARS AND GALAXIES: DESCRIPTIVE ASTRONOMY FOR BEGINNERS, A. E. Fanning. Comprehensive introductory survey of astronomy: the sun, solar system, stars, galaxies, universe, cosmology; up-to-date, including quasars, radio stars, etc. Preface by Prof. Donald Menzel. 24pp. of photographs. 189pp. 5¼ x 8¼.
21680-2 Paperbound $1.75

TEACH YOURSELF CALCULUS, P. Abbott. With a good background in algebra and trig, you can teach yourself calculus with this book. Simple, straightforward introduction to functions of all kinds, integration, differentiation, series, etc. "Students who are beginning to study calculus method will derive great help from this book." Faraday House Journal. 308pp. 20683-1 Clothbound $2.50

TEACH YOURSELF TRIGONOMETRY, P. Abbott. Geometrical foundations, indices and logarithms, ratios, angles, circular measure, etc. are presented in this sound, easy-to-use text. Excellent for the beginner or as a brush up, this text carries the student through the solution of triangles. 204pp. 20682-3 Clothbound $2.50

BASIC MACHINES AND HOW THEY WORK, U. S. Bureau of Naval Personnel. Originally used in U.S. Naval training schools, this book clearly explains the operation of a progression of machines, from the simplest—lever, wheel and axle, inclined plane, wedge, screw—to the most complex—typewriter, internal combustion engine, computer mechanism. Utilizing an approach that requires only an elementary understanding of mathematics, these explanations build logically upon each other and are assisted by over 200 drawings and diagrams. Perfect as a technical school manual or as a self-teaching aid to the layman. 204 figures. Preface. Index. vii + 161pp. 6½ x 9¼. 21709-4 Paperbound $2.50

THE FRIENDLY STARS, Martha Evans Martin. Classic has taught naked-eye observation of stars, planets to hundreds of thousands, still not surpassed for charm, lucidity, adequacy. Completely updated by Professor Donald H. Menzel, Harvard Observatory. 25 illustrations. 16 x 30 chart. x + 147pp. 21099-5 Paperbound $1.50

MUSIC OF THE SPHERES: THE MATERIAL UNIVERSE FROM ATOM TO QUASAR, SIMPLY EXPLAINED, Guy Murchie. Extremely broad, brilliantly written popular account begins with the solar system and reaches to dividing line between matter and nonmatter; latest understandings presented with exceptional clarity. Volume One: Planets, stars, galaxies, cosmology, geology, celestial mechanics, latest astronomical discoveries; Volume Two: Matter, atoms, waves, radiation, relativity, chemical action, heat, nuclear energy, quantum theory, music, light, color, probability, antimatter, antigravity, and similar topics. 319 figures. 1967 (second) edition. Total of xx + 644pp. 21809-0, 21810-4 Two volumes, Paperbound $5.50

OLD-TIME SCHOOLS AND SCHOOL BOOKS, Clifton Johnson. Illustrations and rhymes from early primers, abundant quotations from early textbooks, many anecdotes of school life enliven this study of elementary schools from Puritans to middle 19th century. Introduction by Carl Withers. 234 illustrations. xxxiii + 381pp.
21031-6 Paperbound $3.50

THE PRINCIPLES OF PSYCHOLOGY, William James. The famous long course, complete and unabridged. Stream of thought, time perception, memory, experimental methods—these are only some of the concerns of a work that was years ahead of its time and still valid, interesting, useful. 94 figures. Total of xviii + 1391pp.
20381-6, 20382-4 Two volumes, Paperbound $8.00

THE STRANGE STORY OF THE QUANTUM, Banesh Hoffmann. Non-mathematical but thorough explanation of work of Planck, Einstein, Bohr, Pauli, de Broglie, Schrödinger, Heisenberg, Dirac, Feynman, etc. No technical background needed. "Of books attempting such an account, this is the best," Henry Margenau, Yale. 40-page "Postscript 1959." xii + 285pp. 20518-5 Paperbound $2.00

THE RISE OF THE NEW PHYSICS, A. d'Abro. Most thorough explanation in print of central core of mathematical physics, both classical and modern; from Newton to Dirac and Heisenberg. Both history and exposition; philosophy of science, causality, explanations of higher mathematics, analytical mechanics, electromagnetism, thermodynamics, phase rule, special and general relativity, matrices. No higher mathematics needed to follow exposition, though treatment is elementary to intermediate in level. Recommended to serious student who wishes verbal understanding. 97 illustrations. xvii + 982pp. 20003-5, 20004-3 Two volumes, Paperbound $6.00

GREAT IDEAS OF OPERATIONS RESEARCH, Jagjit Singh. Easily followed non-technical explanation of mathematical tools, aims, results: statistics, linear programming, game theory, queueing theory, Monte Carlo simulation, etc. Uses only elementary mathematics. Many case studies, several analyzed in detail. Clarity, breadth make this excellent for specialist in another field who wishes background. 41 figures. x + 228pp. 21886-4 Paperbound $2.50

GREAT IDEAS OF MODERN MATHEMATICS: THEIR NATURE AND USE, Jagjit Singh. Internationally famous expositor, winner of Unesco's Kalinga Award for science popularization explains verbally such topics as differential equations, matrices, groups, sets, transformations, mathematical logic and other important modern mathematics, as well as use in physics, astrophysics, and similar fields. Superb exposition for layman, scientist in other areas. viii + 312pp.
20587-8 Paperbound $2.50

GREAT IDEAS IN INFORMATION THEORY, LANGUAGE AND CYBERNETICS, Jagjit Singh. The analog and digital computers, how they work, how they are like and unlike the human brain, the men who developed them, their future applications, computer terminology. An essential book for today, even for readers with little math. Some mathematical demonstrations included for more advanced readers. 118 figures. Tables. ix + 338pp. 21694-2 Paperbound $2.50

CHANCE, LUCK AND STATISTICS, Horace C. Levinson. Non-mathematical presentation of fundamentals of probability theory and science of statistics and their applications. Games of chance, betting odds, misuse of statistics, normal and skew distributions, birth rates, stock speculation, insurance. Enlarged edition. Formerly "The Science of Chance." xiii + 357pp. 21007-3 Paperbound $2.50

ADVENTURES OF AN AFRICAN SLAVER, Theodore Canot. Edited by Brantz Mayer. A detailed portrayal of slavery and the slave trade, 1820-1840. Canot, an established trader along the African coast, describes the slave economy of the African kingdoms, the treatment of captured negroes, the extensive journeys in the interior to gather slaves, slave revolts and their suppression, harems, bribes, and much more. Full and unabridged republication of 1854 edition. Introduction by Malcom Cowley. 16 illustrations. xvii + 448pp. 22456-2 Paperbound $3.50

MY BONDAGE AND MY FREEDOM, Frederick Douglass. Born and brought up in slavery, Douglass witnessed its horrors and experienced its cruelties, but went on to become one of the most outspoken forces in the American anti-slavery movement. Considered the best of his autobiographies, this book graphically describes the inhuman treatment of slaves, its effects on slave owners and slave families, and how Douglass's determination led him to a new life. Unaltered reprint of 1st (1855) edition. xxxii + 464pp. 22457-0 Paperbound $2.50

THE INDIANS' BOOK, recorded and edited by Natalie Curtis. Lore, music, narratives, dozens of drawings by Indians themselves from an authoritative and important survey of native culture among Plains, Southwestern, Lake and Pueblo Indians. Standard work in popular ethnomusicology. 149 songs in full notation. 23 drawings, 23 photos. xxxi + 584pp. 6⅝ x 9⅜. 21939-9 Paperbound $4.50

DICTIONARY OF AMERICAN PORTRAITS, edited by Hayward and Blanche Cirker. 4024 portraits of 4000 most important Americans, colonial days to 1905 (with a few important categories, like Presidents, to present). Pioneers, explorers, colonial figures, U. S. officials, politicians, writers, military and naval men, scientists, inventors, manufacturers, jurists, actors, historians, educators, notorious figures, Indian chiefs, etc. All authentic contemporary likenesses. The only work of its kind in existence; supplements all biographical sources for libraries. Indispensable to anyone working with American history. 8,000-item classified index, finding lists, other aids. xiv + 756pp. 9¼ x 12¾. 21823-6 Clothbound $30.00

TRITTON'S GUIDE TO BETTER WINE AND BEER MAKING FOR BEGINNERS, S. M. Tritton. All you need to know to make family-sized quantities of over 100 types of grape, fruit, herb and vegetable wines; as well as beers, mead, cider, etc. Complete recipes, advice as to equipment, procedures such as fermenting, bottling, and storing wines. Recipes given in British, U. S., and metric measures. Accompanying booklet lists sources in U. S. A. where ingredients may be bought, and additional information. 11 illustrations. 157pp. 5⅝ x 8⅛.
(USO) 22090-7 Clothbound $3.50

GARDENING WITH HERBS FOR FLAVOR AND FRAGRANCE, Helen M. Fox. How to grow herbs in your own garden, how to use them in your cooking (over 55 recipes included), legends and myths associated with each species, uses in medicine, perfumes, etc.—these are elements of one of the few books written especially for American herb fanciers. Guides you step-by-step from soil preparation to harvesting and storage for each type of herb. 12 drawings by Louise Mansfield. xiv + 334pp. 22540-2 Paperbound $2.50

JIM WHITEWOLF: THE LIFE OF A KIOWA APACHE INDIAN, Charles S. Brant, editor. Spans transition between native life and acculturation period, 1880 on. Kiowa culture, personal life pattern, religion and the supernatural, the Ghost Dance, breakdown in the White Man's world, similar material. 1 map. xii + 144pp.
22015-X Paperbound $1.75

THE NATIVE TRIBES OF CENTRAL AUSTRALIA, Baldwin Spencer and F. J. Gillen. Basic book in anthropology, devoted to full coverage of the Arunta and Warramunga tribes; the source for knowledge about kinship systems, material and social culture, religion, etc. Still unsurpassed. 121 photographs, 89 drawings. xviii + 669pp.
21775-2 Paperbound $5.00

MALAY MAGIC, Walter W. Skeat. Classic (1900); still the definitive work on the folklore and popular religion of the Malay peninsula. Describes marriage rites, birth spirits and ceremonies, medicine, dances, games, war and weapons, etc. Extensive quotes from original sources, many magic charms translated into English. 35 illustrations. Preface by Charles Otto Blagden. xxiv + 685pp.
21760-4 Paperbound $4.00

HEAVENS ON EARTH: UTOPIAN COMMUNITIES IN AMERICA, 1680-1880, Mark Holloway. The finest nontechnical account of American utopias, from the early Woman in the Wilderness, Ephrata, Rappites to the enormous mid 19th-century efflorescence; Shakers, New Harmony, Equity Stores, Fourier's Phalanxes, Oneida, Amana, Fruitlands, etc. "Entertaining and very instructive." *Times Literary Supplement.* 15 illustrations. 246pp.
21593-8 Paperbound $2.00

LONDON LABOUR AND THE LONDON POOR, Henry Mayhew. Earliest (c. 1850) sociological study in English, describing myriad subcultures of London poor. Particularly remarkable for the thousands of pages of direct testimony taken from the lips of London prostitutes, thieves, beggars, street sellers, chimney-sweepers, street-musicians, "mudlarks," "pure-finders," rag-gatherers, "running-patterers," dock laborers, cab-men, and hundreds of others, quoted directly in this massive work. An extraordinarily vital picture of London emerges. 110 illustrations. Total of lxxvi + 1951pp. 6⅝ x 10.
21934-8, 21935-6, 21936-4, 21937-2 Four volumes, Paperbound $16.00

HISTORY OF THE LATER ROMAN EMPIRE, J. B. Bury. Eloquent, detailed reconstruction of Western and Byzantine Roman Empire by a major historian, from the death of Theodosius I (395 A.D.) to the death of Justinian (565). Extensive quotations from contemporary sources; full coverage of important Roman and foreign figures of the time. xxxiv + 965pp. 20398-0, 20399-9 Two volumes, Paperbound $7.00

AN INTELLECTUAL AND CULTURAL HISTORY OF THE WESTERN WORLD, Harry Elmer Barnes. Monumental study, tracing the development of the accomplishments that make up human culture. Every aspect of man's achievement surveyed from its origins in the Paleolithic to the present day (1964); social structures, ideas, economic systems, art, literature, technology, mathematics, the sciences, medicine, religion, jurisprudence, etc. Evaluations of the contributions of scores of great men. 1964 edition, revised and edited by scholars in the many fields represented. Total of xxix + 1381pp. 21275-0, 21276-9, 21277-7 Three volumes, Paperbound $10.50

AMERICAN FOOD AND GAME FISHES, David S. Jordan and Barton W. Evermann. Definitive source of information, detailed and accurate enough to enable the sportsman and nature lover to identify conclusively some 1,000 species and sub-species of North American fish, sought for food or sport. Coverage of range, physiology, habits, life history, food value. Best methods of capture, interest to the angler, advice on bait, fly-fishing, etc. 338 drawings and photographs. 1 + 574pp. 6⅝ x 9⅜.
22383-1 Paperbound $4.50

THE FROG BOOK, Mary C. Dickerson. Complete with extensive finding keys, over 300 photographs, and an introduction to the general biology of frogs and toads, this is the classic non-technical study of Northeastern and Central species. 58 species; 290 photographs and 16 color plates. xvii + 253pp.
21973-9 Paperbound $4.00

THE MOTH BOOK: A GUIDE TO THE MOTHS OF NORTH AMERICA, William J. Holland. Classical study, eagerly sought after and used for the past 60 years. Clear identification manual to more than 2,000 different moths, largest manual in existence. General information about moths, capturing, mounting, classifying, etc., followed by species by species descriptions. 263 illustrations plus 48 color plates show almost every species, full size. 1968 edition, preface, nomenclature changes by A. E. Brower. xxiv + 479pp. of text. 6½ x 9¼.
21948-8 Paperbound $5.00

THE SEA-BEACH AT EBB-TIDE, Augusta Foote Arnold. Interested amateur can identify hundreds of marine plants and animals on coasts of North America; marine algae; seaweeds; squids; hermit crabs; horse shoe crabs; shrimps; corals; sea anemones; etc. Species descriptions cover: structure; food; reproductive cycle; size; shape; color; habitat; etc. Over 600 drawings. 85 plates. xii + 490pp.
21949-6 Paperbound $3.50

COMMON BIRD SONGS, Donald J. Borror. 33⅓ 12-inch record presents songs of 60 important birds of the eastern United States. A thorough, serious record which provides several examples for each bird, showing different types of song, individual variations, etc. Inestimable identification aid for birdwatcher. 32-page booklet gives text about birds and songs, with illustration for each bird.
21829-5 Record, book, album. Monaural. $2.75

FADS AND FALLACIES IN THE NAME OF SCIENCE, Martin Gardner. Fair, witty appraisal of cranks and quacks of science: Atlantis, Lemuria, hollow earth, flat earth, Velikovsky, orgone energy, Dianetics, flying saucers, Bridey Murphy, food fads, medical fads, perpetual motion, etc. Formerly "In the Name of Science." x + 363pp.
20394-8 Paperbound $2.00

HOAXES, Curtis D. MacDougall. Exhaustive, unbelievably rich account of great hoaxes: Locke's moon hoax, Shakespearean forgeries, sea serpents, Loch Ness monster, Cardiff giant, John Wilkes Booth's mummy, Disumbrationist school of art, dozens more; also journalism, psychology of hoaxing. 54 illustrations. xi + 338pp.
20465-0 Paperbound $2.75

THE PHILOSOPHY OF THE UPANISHADS, Paul Deussen. Clear, detailed statement of upanishadic system of thought, generally considered among best available. History of these works, full exposition of system emergent from them, parallel concepts in the West. Translated by A. S. Geden. xiv + 429pp.

21616-0 Paperbound $3.50

LANGUAGE, TRUTH AND LOGIC, Alfred J. Ayer. Famous, remarkably clear introduction to the Vienna and Cambridge schools of Logical Positivism; function of philosophy, elimination of metaphysical thought, nature of analysis, similar topics. "Wish I had written it myself," Bertrand Russell. 2nd, 1946 edition. 160pp.

20010-8 Paperbound $1.50

THE GUIDE FOR THE PERPLEXED, Moses Maimonides. Great classic of medieval Judaism, major attempt to reconcile revealed religion (Pentateuch, commentaries) and Aristotelian philosophy. Enormously important in all Western thought. Unabridged Friedländer translation. 50-page introduction. lix + 414pp.

(USO) 20351-4 Paperbound $3.50

OCCULT AND SUPERNATURAL PHENOMENA, D. H. Rawcliffe. Full, serious study of the most persistent delusions of mankind: crystal gazing, mediumistic trance, stigmata, lycanthropy, fire walking, dowsing, telepathy, ghosts, ESP, etc., and their relation to common forms of abnormal psychology. Formerly *Illusions and Delusions of the Supernatural and the Occult*. iii + 551pp.   20503-7 Paperbound $3.50

THE EGYPTIAN BOOK OF THE DEAD: THE PAPYRUS OF ANI, E. A. Wallis Budge. Full hieroglyphic text, interlinear transliteration of sounds, word for word translation, then smooth, connected translation; Theban recension. Basic work in Ancient Egyptian civilization; now even more significant than ever for historical importance, dilation of consciousness, etc. clvi + 377pp. 6½ x 9¼.

21866-X Paperbound $3.95

PSYCHOLOGY OF MUSIC, Carl E. Seashore. Basic, thorough survey of everything known about psychology of music up to 1940's; essential reading for psychologists, musicologists. Physical acoustics; auditory apparatus; relationship of physical sound to perceived sound; role of the mind in sorting, altering, suppressing, creating sound sensations; musical learning, testing for ability, absolute pitch, other topics. Records of Caruso, Menuhin analyzed. 88 figures. xix + 408pp.

21851-1 Paperbound $3.50

THE I CHING (THE BOOK OF CHANGES), translated by James Legge. Complete translated text plus appendices by Confucius, of perhaps the most penetrating divination book ever compiled. Indispensable to all study of early Oriental civilizations. 3 plates. xxiii + 448pp.   21062-6 Paperbound $3.00

THE UPANISHADS, translated by Max Müller. Twelve classical upanishads: Chandogya, Kena, Aitareya, Kaushitaki, Isa, Katha, Mundaka, Taittiriyaka, Brhadaranyaka, Svetasvatara, Prasna, Maitriyana. 160-page introduction, analysis by Prof. Müller. Total of 670pp. 20992-X, 20993-8 Two volumes, Paperbound $6.50

VISUAL ILLUSIONS: THEIR CAUSES, CHARACTERISTICS, AND APPLICATIONS, Matthew Luckiesh. Thorough description and discussion of optical illusion, geometric and perspective, particularly; size and shape distortions, illusions of color, of motion; natural illusions; use of illusion in art and magic, industry, etc. Most useful today with op art, also for classical art. Scores of effects illustrated. Introduction by William H. Ittleson. 100 illustrations. xxi + 252pp.

21530-X Paperbound $2.00

A HANDBOOK OF ANATOMY FOR ART STUDENTS, Arthur Thomson. Thorough, virtually exhaustive coverage of skeletal structure, musculature, etc. Full text, supplemented by anatomical diagrams and drawings and by photographs of undraped figures. Unique in its comparison of male and female forms, pointing out differences of contour, texture, form. 211 figures, 40 drawings, 86 photographs. xx + 459pp. 5⅜ x 8⅜.

21163-0 Paperbound $3.50

150 MASTERPIECES OF DRAWING, Selected by Anthony Toney. Full page reproductions of drawings from the early 16th to the end of the 18th century, all beautifully reproduced: Rembrandt, Michelangelo, Dürer, Fragonard, Urs, Graf, Wouwerman, many others. First-rate browsing book, model book for artists. xviii + 150pp. 8⅜ x 11¼.

21032-4 Paperbound $2.50

THE LATER WORK OF AUBREY BEARDSLEY, Aubrey Beardsley. Exotic, erotic, ironic masterpieces in full maturity: Comedy Ballet, Venus and Tannhauser, Pierrot, Lysistrata, Rape of the Lock, Savoy material, Ali Baba, Volpone, etc. This material revolutionized the art world, and is still powerful, fresh, brilliant. With *The Early Work*, all Beardsley's finest work. 174 plates, 2 in color. xiv + 176pp. 8⅛ x 11.

21817-1 Paperbound $3.00

DRAWINGS OF REMBRANDT, Rembrandt van Rijn. Complete reproduction of fabulously rare edition by Lippmann and Hofstede de Groot, completely reedited, updated, improved by Prof. Seymour Slive, Fogg Museum. Portraits, Biblical sketches, landscapes, Oriental types, nudes, episodes from classical mythology—All Rembrandt's fertile genius. Also selection of drawings by his pupils and followers. "Stunning volumes," *Saturday Review*. 550 illustrations. lxxviii + 552pp. 9⅛ x 12¼.

21485-0, 21486-9 Two volumes, Paperbound $10.00

THE DISASTERS OF WAR, Francisco Goya. One of the masterpieces of Western civilization—83 etchings that record Goya's shattering, bitter reaction to the Napoleonic war that swept through Spain after the insurrection of 1808 and to war in general. Reprint of the first edition, with three additional plates from Boston's Museum of Fine Arts. All plates facsimile size. Introduction by Philip Hofer, Fogg Museum. v + 97pp. 9⅜ x 8¼.

21872-4 Paperbound $2.00

GRAPHIC WORKS OF ODILON REDON. Largest collection of Redon's graphic works ever assembled: 172 lithographs, 28 etchings and engravings, 9 drawings. These include some of his most famous works. All the plates from *Odilon Redon: oeuvre graphique complet*, plus additional plates. New introduction and caption translations by Alfred Werner. 209 illustrations. xxvii + 209pp. 9⅛ x 12¼.

21966-8 Paperbound $4.00

INCIDENTS OF TRAVEL IN YUCATAN, John L. Stephens. Classic (1843) exploration of jungles of Yucatan, looking for evidences of Maya civilization. Stephens found many ruins; comments on travel adventures, Mexican and Indian culture. 127 striking illustrations by F. Catherwood. Total of 669 pp.
20926-1, 20927-X Two volumes, Paperbound $5.00

INCIDENTS OF TRAVEL IN CENTRAL AMERICA, CHIAPAS, AND YUCATAN, John L. Stephens. An exciting travel journal and an important classic of archeology. Narrative relates his almost single-handed discovery of the Mayan culture, and exploration of the ruined cities of Copan, Palenque, Utatlan and others; the monuments they dug from the earth, the temples buried in the jungle, the customs of poverty-stricken Indians living a stone's throw from the ruined palaces. 115 drawings by F. Catherwood. Portrait of Stephens. xii + 812pp.
22404-X, 22405-8 Two volumes, Paperbound $6.00

A NEW VOYAGE ROUND THE WORLD, William Dampier. Late 17-century naturalist joined the pirates of the Spanish Main to gather information; remarkably vivid account of buccaneers, pirates; detailed, accurate account of botany, zoology, ethnography of lands visited. Probably the most important early English voyage, enormous implications for British exploration, trade, colonial policy. Also most interesting reading. Argonaut edition, introduction by Sir Albert Gray. New introduction by Percy Adams. 6 plates, 7 illustrations. xlvii + 376pp. 6½ x 9¼.
21900-3 Paperbound $3.00

INTERNATIONAL AIRLINE PHRASE BOOK IN SIX LANGUAGES, Joseph W. Bátor. Important phrases and sentences in English paralleled with French, German, Portuguese, Italian, Spanish equivalents, covering all possible airport-travel situations; created for airline personnel as well as tourist by Language Chief, Pan American Airlines. xiv + 204pp.
22017-6 Paperbound $2.00

STAGE COACH AND TAVERN DAYS, Alice Morse Earle. Detailed, lively account of the early days of taverns; their uses and importance in the social, political and military life; furnishings and decorations; locations; food and drink; tavern signs, etc. Second half covers every aspect of early travel; the roads, coaches, drivers, etc. Nostalgic, charming, packed with fascinating material. 157 illustrations, mostly photographs. xiv + 449pp.
22518-6 Paperbound $4.00

NORSE DISCOVERIES AND EXPLORATIONS IN NORTH AMERICA, Hjalmar R. Holand. The perplexing Kensington Stone, found in Minnesota at the end of the 19th century. Is it a record of a Scandinavian expedition to North America in the 14th century? Or is it one of the most successful hoaxes in history. A scientific detective investigation. Formerly *Westward from Vinland.* 31 photographs, 17 figures. x + 354pp.
22014-1 Paperbound $2.75

A BOOK OF OLD MAPS, compiled and edited by Emerson D. Fite and Archibald Freeman. 74 old maps offer an unusual survey of the discovery, settlement and growth of America down to the close of the Revolutionary war: maps showing Norse settlements in Greenland, the explorations of Columbus, Verrazano, Cabot, Champlain, Joliet, Drake, Hudson, etc., campaigns of Revolutionary war battles, and much more. Each map is accompanied by a brief historical essay. xvi + 299pp. 11 x 13¾.
22084-2 Paperbound $6.00

Two Little Savages; Being the Adventures of Two Boys Who Lived as Indians and What They Learned, Ernest Thompson Seton. Great classic of nature and boyhood provides a vast range of woodlore in most palatable form, a genuinely entertaining story. Two farm boys build a teepee in woods and live in it for a month, working out Indian solutions to living problems, star lore, birds and animals, plants, etc. 293 illustrations. vii + 286pp.

20985-7 Paperbound $2.50

Peter Piper's Practical Principles of Plain & Perfect Pronunciation. Alliterative jingles and tongue-twisters of surprising charm, that made their first appearance in America about 1830. Republished in full with the spirited woodcut illustrations from this earliest American edition. 32pp. $4\frac{1}{2}$ x $6\frac{3}{8}$.

22560-7 Paperbound $1.00

Science Experiments and Amusements for Children, Charles Vivian. 73 easy experiments, requiring only materials found at home or easily available, such as candles, coins, steel wool, etc.; illustrate basic phenomena like vacuum, simple chemical reaction, etc. All safe. Modern, well-planned. Formerly *Science Games for Children*. 102 photos, numerous drawings. 96pp. $6\frac{1}{8}$ x $9\frac{1}{4}$.

21856-2 Paperbound $1.25

An Introduction to Chess Moves and Tactics Simply Explained, Leonard Barden. Informal intermediate introduction, quite strong in explaining reasons for moves. Covers basic material, tactics, important openings, traps, positional play in middle game, end game. Attempts to isolate patterns and recurrent configurations. Formerly *Chess*. 58 figures. 102pp. (USO) 21210-6 Paperbound $1.25

Lasker's Manual of Chess, Dr. Emanuel Lasker. Lasker was not only one of the five great World Champions, he was also one of the ablest expositors, theorists, and analysts. In many ways, his Manual, permeated with his philosophy of battle, filled with keen insights, is one of the greatest works ever written on chess. Filled with analyzed games by the great players. A single-volume library that will profit almost any chess player, beginner or master. 308 diagrams. xli x 349pp.

20640-8 Paperbound $2.75

The Master Book of Mathematical Recreations, Fred Schuh. In opinion of many the finest work ever prepared on mathematical puzzles, stunts, recreations; exhaustively thorough explanations of mathematics involved, analysis of effects, citation of puzzles and games. Mathematics involved is elementary. Translated by F. Göbel. 194 figures. xxiv + 430pp.

22134-2 Paperbound $3.00

Mathematics, Magic and Mystery, Martin Gardner. Puzzle editor for Scientific American explains mathematics behind various mystifying tricks: card tricks, stage "mind reading," coin and match tricks, counting out games, geometric dissections, etc. Probability sets, theory of numbers clearly explained. Also provides more than 400 tricks, guaranteed to work, that you can do. 135 illustrations. xii + 176pp.

20338-2 Paperbound $1.50

f

MATHEMATICAL PUZZLES FOR BEGINNERS AND ENTHUSIASTS, Geoffrey Mott-Smith. 189 puzzles from easy to difficult—involving arithmetic, logic, algebra, properties of digits, probability, etc.—for enjoyment and mental stimulus. Explanation of mathematical principles behind the puzzles. 135 illustrations. viii + 248pp.
20198-8 Paperbound $1.75

PAPER FOLDING FOR BEGINNERS, William D. Murray and Francis J. Rigney. Easiest book on the market, clearest instructions on making interesting, beautiful origami. Sail boats, cups, roosters, frogs that move legs, bonbon boxes, standing birds, etc. 40 projects; more than 275 diagrams and photographs. 94pp.
20713-7 Paperbound $1.00

TRICKS AND GAMES ON THE POOL TABLE, Fred Herrmann. 79 tricks and games— some solitaires, some for two or more players, some competitive games—to entertain you between formal games. Mystifying shots and throws, unusual caroms, tricks involving such props as cork, coins, a hat, etc. Formerly *Fun on the Pool Table*. 77 figures. 95pp.
21814-7 Paperbound $1.00

HAND SHADOWS TO BE THROWN UPON THE WALL: A SERIES OF NOVEL AND AMUSING FIGURES FORMED BY THE HAND, Henry Bursill. Delightful picturebook from great-grandfather's day shows how to make 18 different hand shadows: a bird that flies, duck that quacks, dog that wags his tail, camel, goose, deer, boy, turtle, etc. Only book of its sort. vi + 33pp. 6½ x 9¼. 21779-5 Paperbound $1.00

WHITTLING AND WOODCARVING, E. J. Tangerman. 18th printing of best book on market. "If you can cut a potato you can carve" toys and puzzles, chains, chessmen, caricatures, masks, frames, woodcut blocks, surface patterns, much more. Information on tools, woods, techniques. Also goes into serious wood sculpture from Middle Ages to present, East and West. 464 photos, figures. x + 293pp.
20965-2 Paperbound $2.00

HISTORY OF PHILOSOPHY, Julián Marias. Possibly the clearest, most easily followed, best planned, most useful one-volume history of philosophy on the market; neither skimpy nor overfull. Full details on system of every major philosopher and dozens of less important thinkers from pre-Socratics up to Existentialism and later. Strong on many European figures usually omitted. Has gone through dozens of editions in Europe. 1966 edition, translated by Stanley Appelbaum and Clarence Strowbridge. xviii + 505pp. 21739-6 Paperbound $3.00

YOGA: A SCIENTIFIC EVALUATION, Kovoor T. Behanan. Scientific but non-technical study of physiological results of yoga exercises; done under auspices of Yale U. Relations to Indian thought, to psychoanalysis, etc. 16 photos. xxiii + 270pp.
20505-3 Paperbound $2.50

Prices subject to change without notice.
Available at your book dealer or write for free catalogue to Dept. GI, Dover Publications, Inc., 180 Varick St., N. Y., N. Y. 10014. Dover publishes more than 150 books each year on science, elementary and advanced mathematics, biology, music, art, literary history, social sciences and other areas.